Barbara A. Gilchrest, Jean Krutmann (Eds.)

Skin Aging

Barbara A. Gilchrest   Jean Krutmann (Eds.)

Skin Aging

With 72 Figures and 24 Tables

 Springer

Prof. Dr. Barbara A. Gilchrest
Department of Dermatology
Boston University School of Medicine
609 Albany Street
Boston, MA 02118
USA

Prof. Dr. Jean Krutmann
Institut für Umweltmedizinische Forschung (IUF)
at the Heinrich-Heine-University
Düsseldorf gGmbH
Auf'm Hennekamp 50
40225 Düsseldorf
Germany

Library of Congress Control Number: 2005933930

ISBN-10 3-540-24443-3   Springer Berlin Heidelberg New York
ISBN-13 978-3-540-24443-1   Springer Berlin Heidelberg New York

Springer is a part of Springer Science + Business Media
springer.com

© Springer-Verlag Berlin Heidelberg 2006
Printed in Germany

Editor: Marion Philipp, Heidelberg
Desk Editor: Ellen Blasig, Heidelberg
Cover design: Frido Steinen-Broo, eStudio Calamar, Spain
Typesetting: Satz-Druck-Service, Leimen
Production: LE-TEX Jelonek, Schmidt & Vöckler GbR, Leipzig

Printed on acid-free paper   24/3180 – YL – 5 4 3 2 1   SPIN 12438501 dedicated to Claudia Billmann-Krutmann, M.D.

Foreword

A decade ago Barbara Gilchrest published a landmark volume titled Photodamage consisting of a mere 285 pages and somewhat more than a dozen chapters.

I wrote in the foreword that this field of study "was in its adolescence, a thriving enterprise, delivering new discoveries every month."

The present volume by Gilchrest and Krutmann, titled Skin Aging, is a greatly expanded outgrowth of the original, which is at least double in size and with double the number of authors and chapters. It stands as a testimony to the enormous advances in knowledge that have occurred in ten years, not only with an avalanche of publications, but more profound insights made possible by molecular biology, genomics, and immunohistochemistry, among other rapidly developing disciplines. Indeed, photoaging owes its advances to its multidisciplinary structure.

This volume strikes a balance between the basic science of photobiology with an emphasis on photodamage and clinical interventions to combat and correct the destructive effects of solar radiation. One of the main themes is the prevention of photodamage, a timely focus, reflecting a shift in the traditional biomedical paradigm, which now exalts prevention over treatment. After all, treatment really means failure of prevention. Happily we can now claim that photoaging is not only not inevitable, but is entirely preventable by following a few simple rules. People who look much older than their chronologic age, previously termed premature aging, have simply failed to heed our advice out of ignorance or perversity!

This text is not only an invaluable reference source for a diverse readership of basic scientists, in academia and industry, but perhaps more importantly, a training manual for public health authorities and educators who seek to inform a health conscious public about the best ways to prevent photoaging in our sun-worshipping culture, preserving an attractive, youthful appearance well into old age.

It is more than a hundred years since Unna from Germany and Dubreiulh from France explicitly described "farmers' and sailors' skin" which they correctly attributed to excessive exposure to sunlight.

Curiously these warnings fell on deaf ears for almost another 75 years until dermatologists, caring for an epidemic of aging persons concerned with a deterioration in appearance, which coincided with an epidemic of skin cancers, reignited the subject, fostering serious investigations into the basic biology of photoaging. The last 25 years of the 20th century witnessed an avalanche of publications, relating to all aspects of the subject, more than a thousand papers originating from all corners of the world.

Despite energetic campaigns by national dermatological organizations and constant exhortations by the lay press and mass media, we have largely failed to persuade the public to follow our advice. The proportion of older people showing the dreaded stigmata of wrinkles, blotches, sags, laxity and cancers has continued to grow, fueling a huge increase in the number of physicians practicing cosmetic surgery. This book is an invaluable resource for dermatosurgeons who would be well advised to know the basic science principles that underlie their specialty.

The epidemic of non-melanoma skin cancers, basal cell and squamous cell cancers, continues unabated. It is shocking that the annual incidence of these cancers is approximately equal to the incidence of all other forms of visceral cancers–lung, colon, and cervical–combined! It is estimated that 1 of 70 white Americans will experience a malignant melanoma in their life-time, in which heedless exposure to solar radiation is the predominant etiologic agent.

People are living longer, on average 35 years more than in Unna's time. They also have more wealth and more leisure time for recreational activities and vacations under the sun. Dermatologists have initiated national campaigns to warn against the dangers of the fashionable belief that a deep tan makes one more attractive and happier. Whatever the supposed short-term social benefits, the ultimate long-term effect is the repulsive appearance of wrinkled facial skin, replete with bags and sags, blotchy dyspigmentations, tumorous growths, wrinkles, yellowing, poor texture, clogged pores, and dry-roughness, which certainly do not add to the quality of life of the elderly. Despite the warnings that there is no safe way to obtain a deep tan, tanning salons are thriving, increasing steadily, and amount to a multi-billion dollar business in the USA alone, with few regulations to limit the harm.

In the face of these foibles and follies, photobiologists continue to make surprising discoveries of the powerful capacity of solar radiation to damage unprotected skin. Even a single minimum erythema dose (1 MED) can damage the dermal matrix. One sunburn dose damages the matrix forever. The skin has a long memory!

In biblical Genesis we learn that God created light and declared it to be good, an unassailable truism in the case of visible light. The deity did not take into account ultraviolet light, which is of course invisible and should be called by its proper name, ultraviolet radiation, not ultraviolet light.

Like true missionaries, photobiologists have properly thundered against the multiple dangers of sunlight, but this zeal may need to be balanced against the known beneficial effects of sunlight, a subject not neglected in this volume. For example, sunlight stimulates the epidermis to synthesize pre-vitamin D3, whose absence in the diet leads to osteoporosis and fractures, a major problem among the elderly. Sunlight clearly induces a feeling of euphoria and is extremely beneficial in moderating seasonal affective disorders.

Light can be harnessed by a variety of new technologies, subsumed under the general term of phototherapy, to treat effectively a number of chronic skin disorders such as acne, psoriasis and even cutaneous cancers. These techniques employ a growing variety of ingenious ablative and non-ablative lasers at various wavelengths, frequencies and pulse durations to achieve desired benefits, sometimes abetted by photosensitizing drugs to generate oxidative free radicals to destroy tumors.

These salubrious developments are based solely on science, not empiricism as in the past. It is impossible to exaggerate the extent to which these novel phototherapeutic interventions have been brought to fruition based on the basic understanding of photobiologic principles, amply documented by the international authorities who have contributed to this comprehensive text.

Other developments which hover on the horizon open up novel which approaches for exploiting photobiologic principles to effectively treat a variety of unrelated chronic skin disorders, including potentially lethal ones such as T-cell lymphomas. One fascinating example of things to come relates to the repair of photodamaged DNA. The inability to remove or repair DNA damage induced by sunlight inevitably promotes tumor progression, culminating in a variety of cutaneous tumors, as strikingly illustrated in the devastating hereditable disorder xeroderma pigmentosum. This death-dealing disease results from the lack of specific repair enzymes that enable photodamaged DNA to persist throughout many cell replications, ending in carcinogenesis. The solution is to develop topical systems that can deliver these repair enzymes to the viable epidermis, penetrating the horny layer barrier without disrupting its structure. This remarkable feat has already been accomplished in preliminary studies, using liposomes loaded with repair enzymes.

Finally, a rapidly aging population has created an enormous need for so-called anti-aging remedies to correct the ravages of photodamaged skin. These include scores of oral supplements, hundreds of topicals containing every conceivable "active" ingredient–vitamins, hormones, minerals, macronutrients, free-radical scavengers, anti-oxidants, oriental herbs, ad infinitum.

This galaxy of offerings, all of them unregulated and with no requirement to substantiate claims of efficacy, are the modern equivalent of snake oil, based mainly on advertising hype and unabashed huckstering.

In the face of this maniacal onslaught, aimed mostly at women, there are indeed topicals that possess scientifically verified anti-aging effects for correcting wrinkles, dyspigmentations, laxity, etc. Topical retinoids are the oldest and prominent among these credible pharmacological agents, while others are already in the pipeline.

It is the high duty of the photoaging fraternity to educate the public regarding these phantasmagorical, preposterous products, now a business worth 39 billion dollars annually, which is equivalent to the entire budget of the National Institutes of Health! This text does not duck its responsibility to take up this controversial issue and to provide guidelines of what is credible and what is fakery.

In my coda to the forward of Gilchrest's volume ten years ago, I declared that the future of the specialty of photoaging was radiant (meaning bright). That future is now here. The specialty can now be also eulogized as brilliant (meaning distinguished, beyond compare).

Perhaps we are in reach of Ponce de Leon's dream of a fountain of youth, at least preserving the attractive appearance of youth into old age, based of course on science.

Albert M. Kligman, M.D., Ph.D.

Preface

Life expectancy has risen continuously in developed societies, yet the mystery of aging remains largely unresolved. As a consequence, the prevalence of mental and physical disability and diseases related to old age has increased steeply. To prevent this trend from continuing over the next decades and eventually destabilizing our health-care systems, we must find ways to promote successful, healthy aging. This requires (1) identification of aging mechanisms at a molecular level, (2) reduction of their impact over time on organs and organisms, (3) detection of individual genetic and environmental contributions of these aging mechanisms, and (4) development of diagnostic tools and specific strategies for prevention, regeneration, and compensation to delay unwanted age-associated changes.

Among all the organs, aging of the skin is of particular importance because it affects human health, it has a strong social impact due to its visibility, and it represents an ideal model organ for aging research because of its accessibility and well-studied baseline cell–cell and cell–matrix interactions.

This monograph attempts to provide an up-to-date overview regarding all aspects of skin aging. It includes in-depth discussions of the molecular basis as well as concepts propagated for the diagnosis, treatment, and prevention of skin aging.

The explosion of knowledge in this field over even the past decade is remarkable. To capture the depth and breadth of this learning, we have recruited leading experts from multiple sub-disciplines. All authors are internationally recognized, and we are grateful for their excellent contributions. We hope that Skin Aging will serve you well as a state-of-the-art reference and will further stimulate your interest in this fascinating area.

Barbara A. Gilchrest   Boston, MA, USA
Jean Krutmann   Düsseldorf, Germany
Spring 2006

Contents

List of Contributors

Farrukh Afaq
Department of Dermatology,
University of Wisconsin,
Medical Science Center,
1300 University Avenue,
Madison, WI 53706, USA

Chantal O. Barland
Department of Dermatology,
University of California,
San Francisco, CA, USA

Christian Beier
Department of Dermatology,
J.W. Goethe University Hospital,
Theodor-Stern-Kai 7,
60590 Frankfurt am Main, Germany

Peter M. Elias
Dermatology Service (190),
Veterans Affairs Medical Center,
4150 Clement Street,
San Francisco, CA 94121, USA

Gary J. Fisher
Department of Dermatology,
University of Michigan,
Ann Arbor, MI, USA

Ruby Ghadially
Department of Dermatology,
University of California,
San Francisco, CA, USA

Barbara A. Gilchrest
Department of Dermatology,
Boston University School of Medicine,
609 Albany Street,
Boston, MA 02118, USA

Glenda K. Hall
Department of Dermatology,
Boston University School of Medicine,
609 Albany Street,
Boston, MA 02118, USA

Roland Kaufmann
Department of Dermatology,
J.W. Goethe University Hospital,
Theodor-Stern-Kai 7,
60590 Frankfurt am Main, Germany

Ursula Krämer
Environmental Health Research Institute
(IUF),
Auf'm Hennekamp 50,
40225 Düsseldorf, Germany

Niels C. Krejci-Papa
Department of Dermatology,
Boston University School of Medicine,
609 Albany Street,
Boston, MA 02118, USA

Jean Krutmann
Environmental Health Research Institute
(IUF),
Auf'm Hennekamp 50,
40225 Düsseldorf, Germany

Robert C. Langdon
Department of Dermatology,
Yale University School of Medicine,
New Haven, CT, USA

Evgenia Makrantonaki
Department of Dermatology,
Charité University Medicine Berlin,
Fabeckstrasse 60-62,
14195 Berlin, Germany

Akimichi Morita
Department of Geriatric and Environmental
Dermatology, Nagoya City University Graduate
School of Medical Sciences,
Nagoya 467-8601, Japan

Frederique Morizot
Centre de Recherches et d'Investigations
Épidermiques et Sensorielles (CE.R.I.E.S.),
Neuilly, France

Hasan Mukhtar
Department of Dermatology,
University of Wisconsin,
Medical Science Center,
1300 University Avenue,
Madison, WI 53706, USA

Tania J. Phillips
Department of Dermatology,
Boston University School of Medicine,
609 Albany Street,
Boston, MA 02118, USA

Laure Rittié
Department of Dermatology,
University of Michigan,
Ann Arbor, MI, USA

Stefan M. Schieke
Cardiovascular Branch, National Heart,
Lung and Blood Institute, NIH,
Bethesda, MD, USA

Tamara Schikowski
Environmental Health Research Institute
(IUF),
Auf'm Hennekamp 50,
40225 Düsseldorf, Germany

Peter Schroeder
Environmental Health Research Institute
(IUF),
Auf'm Hennekamp 50,
40225 Düsseldorf, Germany

Nanna Schürer
Department of Dermatology,
University of Osnabrück,
Sedanstrasse 115,
49076 Osnabrück, Germany

Helmut Sies
Institute of Biochemistry
and Molecular Biology I,
Heinrich-Heine University Düsseldorf,
40001 Düsseldorf, Germany

Wilhelm Stahl
Institute of Biochemistry
and Molecular Biology I,
Heinrich-Heine University Düsseldorf,
40001 Düsseldorf, Germany

Rolf-Markus Szeimies
Department of Dermatology,
Regensburg University Hospital,
Franz-Josef-Strauss-Allee 11,
93053 Regensburg, Germany

Jens Thiele
Department of Dermatology,
Northwestern University School of Medicine,
Chicago, IL, USA

Dany J. Touma
Moudabber Center, Kaslik, Lebanon;
Hazmieh International Medical Center,
Beirut, Lebanon

Erwin Tschachler
Department of Dermatology,
University of Vienna Medical School,
Waehringer Gürtel 18-20,
1090 Vienna, Austria

John J. Voorhees
Department of Dermatology,
University of Michigan,
Ann Arbor, MI, USA

Deon Wolpowitz
Department of Dermatology,
Boston University School of Medicine,
609 Albany St,
Boston, MA 02115, USA

Mina Yaar
Department of Dermatology,
Boston University School of Medicine,
609 Albany Street,
Boston, MA 02118, USA

Daniel Yarosh
AGI Dermatics Inc.,
Freeport, NY 11520, USA

Christos C. Zouboulis
Departments of Dermatology and
Immunology, Dessau Medical Center,
Auenweg 38,
06847 Dessau, Germany
and
Charité Universitätsmedizin Berlin,
Campus Benjamin Franklin,
Berlin, Germany

Chapter 1

Recent Demographic Changes and Consequences for Dermatology

1

Ursula Krämer, Tamara Schikowski

Contents

1.1 Introduction

In many countries of the world a demographic transition is happening. This transformation involves aging of the world population, and migration of people into different parts of the world. In times of globalization people migrate from less fortunate countries to the developed world. Major advances in the medical field have led to a significant increase in life expectancy throughout the 20th century. Aging is enhanced by an additional decrease in birth rates in many countries of the world. This demographic transition is dynamic and results in an increased demand on the health sector with questions arising in relation to health-care provision and health needs of the elderly and the moving population. The demographic change has created a challenge for the health-care system and society at large [3].

One major consequence of these changes is the world-wide increase in chronic diseases, which affect older people disproportionately and contribute to disabilities and quality of life. The increasingly aging population and the increased numbers of migrants in the so-called developed world also results in a challenge for dermatologists. The problems of geriatric dermatology and skin disorders which are found in some ethnic groups as well as the demands for cosmetic dermatology are evolving and set out new dimensions.

Not only does human skin undergo chronological changes, but the cumulative exposure to environmental factors (UV irradiation) during a longer life can contribute to skin damage and skin disease. The incidence of skin cancer has dramatically increased independently of age in the past 20 years, mainly as a result of lifestyle changes in combination with increased sun exposure [9].

1.2 Demographic Changes and the Implications for the Population

1.2.1 Life Expectancy Worldwide

- Life expectancy is increasing worldwide.
- In most countries women on average live 7 years longer than men.

The reduction in mortality for every given age group worldwide has led to an enormously increased life expectancy. The reduction in infant and child mortality can be seen as the most significant contributor to this increase. The survival of people until old age is a direct result of the advances in medical and biological sciences.

In the least developed countries life expectancy at birth was 35 years in 1950 and will increase to 65 years in 2050. In the industrialized countries life expectancy was 65 years in 1950 and will increase to 82 years in 2050. Today it is 76 years in these countries. The statistic is similar in all developed nations, with the highest life expectancy in Hong Kong and Japan. The world population was 2.5 billion in 1950 and will increase to 8.9 billion in 2050.

The very old age group (aged 80 years and older, 80+) is growing in numbers faster than any other age group. Today 69 million of the world's population are aged 80+. In 2050 this age group will comprise 377 million. Figures from Germany also show that the 80+ population shows the highest rate of increase since 1960. The number of Germans aged 80+ was 1.2 million in 1960 and had grown to 2.9 million by 1998. Projections show a substantial increase to 4 million by 2010 and to 5.3 million by 2020.

A gender difference in mortality can be observed with older women exceeding older men in the population worldwide. In most parts of the world women live longer than men by an average of 7 years. The life expectancy for girls at birth in 1997 was at least 80 years in most developed countries. Thus, health and socioeconomic problems of aging are largely problems of elderly women, not only in developed countries but also in developing countries. In 2025, 71% of all women aged 60 years and over will live in developing countries, in contrast to 1997 when 58% of women aged 60 years and over lived in these countries.

1.2.2 Aging Worldwide

- Aging has started in all populations worldwide.
- The percentage of people aged 60 years and over will have doubled or tripled by 2050.

Figure 1.1 shows the age distributions of different population groups in the world. A distribution with a pagoda-like form has characterized all populations over thousands of years. Today it can only be seen in the least-developed countries. Only here are children the most numerous and elderly individuals a small fraction. Aging of a population not only means an increase in the number of old people, as described above, but also an increase in the proportion of elderly in the population. The demographic transition from high birth rates and high mortality in populations usually follow four phases:

- High birth and mortality rates prevail, a pattern that today can be found only in the least developed countries.

- Infant mortality decreases and life expectancy increases, leading to a rapid population increase (in Europe this was true in the 19th century with the beginning of industrialization).

- Birth rate decreases and the average age in the population increases (in Europe at the beginning of the 20th century).

- A new balance of low birth rate and low mortality is reached.

As can be seen in Fig. 1.1, all populations of the world are somewhere in this process. In Europe between 1965 and 1975 a second decrease in birth rates started, beginning in the northern parts of Europe, and then reaching the southern parts and now the eastern parts. This second decrease in birth rates was accompanied

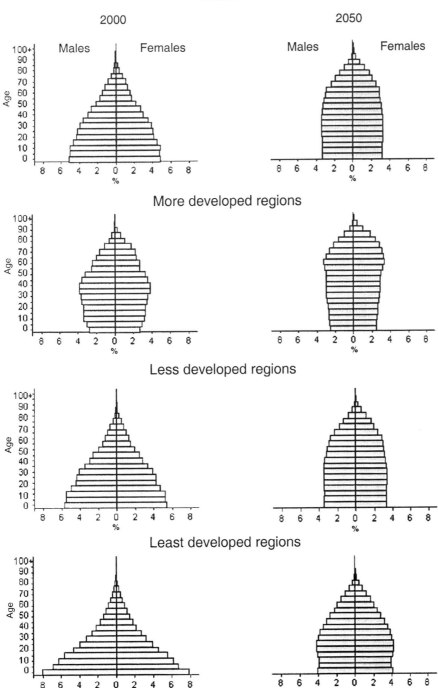

Fig. 1.1. Demographic changes in the population world wide (from: The sex and age distribution of the world population, by the Population Division, Department of Economy, © 2005 United Nations, Reprint with permission of the publisher)

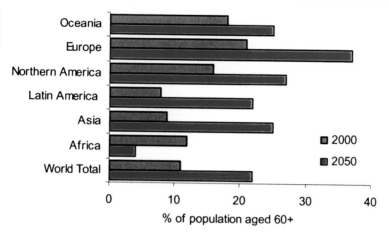

Fig. 1.2. Percentage increase in those aged 60 years and over by region, 2000–2050

by a still-increasing life expectancy leading to a dramatic increase in the elderly population. No steady-state has been reached. The population will decrease in some European countries: in Germany from 82 million in 2000 to about 70 million in 2050. Since this will be the result only of a low birth rate, the percentage of older people will increase. In the year 2000, 17% of the population in Germany were older than 65 years; in 2050 this will have changed to 29%.

Although the proportion of elderly is smaller in most developing countries, it is nevertheless increasing steadily. Current projections indicate that during the first half of this century the number of people in the world aged 60 years and over will nearly double from 231 million to 395 million. The largest increase will occur in developing countries, from 374 million in 2000 to 1.6 billion in 2050 [21] (Fig. 1.2), resulting in enormous socioeconomic and political impacts in all countries.

1.3 The Impact of Aging on Health and Society

- Old age dependency ratio is decreasing dramatically.
- Implementation of preventative strategies is important.

The transition from high birth and death rates to low birth and death rates has consequences for political, socioeconomic and health economic decisions worldwide. Due to the decline in fertility, many industrialized nations are approaching the point where old people outnumber children. This is a particular problem for countries such as Germany and Italy where the populations are decreasing. A measure that describes aging of a population and hints at the social consequences is the "old-age dependency ratio". This is a measure of the numbers of young (working age) individuals (aged 20–64 years) available to support the old (aged 65 years and over). In Germany, for example, this ratio was 10.2 in 1910, 6.2 in 1950 and 3.7 in 2000, and is expected to be 1.8 in 2050.

While medical advances promise enhanced quality of life and independence in old age, current predictions suggest that there will be economic and social inequality for the aging population. With the aging of the population has come the possibility of a longer life-time exposure to various toxic agents, chemical irritants and environmental factors such as sun exposure. The cumulative damage of these agents has profound effects on the health of the elderly. The increasing burden of chronic diseases, injuries and disabilities is a major concern. The morbidities of the aging population put enormous strain on health-care resources.

Not only do the elderly live longer, but they also have higher expectations of quality of life which need to be addressed. The increasing number of aged people puts pressure on the

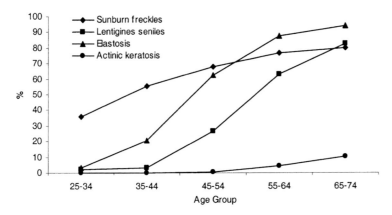

Fig. 1.3. Prevalence of age-related skin changes in different adult age groups in Augsburg, Germany

health system and the medical and social services. The health-care costs for persons aged 65 years and over are growing rapidly and with this growth is a growing need for long-term care. This need is connected to public as well as to personal assets. Figures from the United States show that the costs for nursing homes and home care doubled between 1990 and 2001 [6].

The implementation of preventive strategies to improve the health and wellbeing of the aging population is essential. One major role of modern dermatology is to implement preventive strategies to reduce the prevalence of skin diseases and cancers in the elderly.

1.4 Dermatological Problems of the Aged

1.4.1 Age-Related Skin Changes

- Skin changes which affect appearance increase with age.

The skin is a complex and dynamic organ that shows the most obvious signs of aging. It is in direct contact with the environment and hence undergoes aging also as a consequence of environmental damage. These two processes which affect the skin are clinically and biologically different, and are distinguished as intrinsic and extrinsic aging.

Intrinsically and chronologically aged skin appears to be thinner, more lax and more finely lined. The skin becomes more vulnerable with

age as many naturally protective functions decrease. The so-called biological clock affects the skin in the same way as it affects the internal organs: tissues show slow irreversible degeneration. A natural decline in the functions of the skin manifests clinically as physical findings of decreased turgor, thinning, dryness and ecchymosis [4, 7].

The causes of chronological skin aging are less clear than those of the second type of aging so-called extrinsic aging or photoaging. Photoaging is a result of exposure to environmental elements in particular UV radiation. The affected skin shows a variety of clinical manifestations, among them sunburn freckles, lentigines seniles, elastosis and actinic keratoses. Figure 1.3 depicts the results of a study in Augsburg, Germany [15], one of the few actual determinations of the prevalence of these skin changes in different age groups.

1.4.2 Age-Related Skin Diseases

- Skin diseases which have to be treated by dermatologists increase with age.
- Including skin cancer.

Many protective functions of the skin are impaired as aging progresses, and as a consequence the skin becomes more vulnerable. The burden of dermatological disease in the older population is significant and often underestimated [12]. Many skin disorders are seen more often in elderly patients, including pruritus,

Table 1.1. Pattern of skin diseases in four countries (as cited by Liao et al [14]) (percent of diagnoses)

	Canada (Ottawa) Adam et al 1985–86 (n=326)	Japan (Tokyo) Yamamoto et al 1981–89 (n=10,113)	Singapore Yap et al 1990 (n=2,571)	Taiwan (Taipei) Liao et al 1993–99 (n=16,924)
Dermatitis	16.3	33.7	35.3	58.7
Benign tumours	13.8	4.6	4.1	12.8
Actinic keratosis	24.9	0.3	0.0	0.5
Malignant tumours	12.6	1.0	0.7	2.1
Fungal infection	3.4	16.8	2.6	38.0
Pruritus	1.2	7.5	1.7	14.2

seborrhoeic dermatitis and xerosis [13]. Some disorders are far more common in old age, such as bullous pemphigoid, skin tumours and leg ulcerations. Furthermore, old people often suffer from hyperkeratotic lesions of their feet [19]. Cutaneous manifestations of endocrine and metabolic diseases are seen more often in the elderly because the underlying diseases are more prevalent [16].

In a study by Smith et al. of Australian nursing home patients aged 80 years and over, 29.5% of all patients were affected by xerosis, 22.5% suffered from onychomycosis and 8.9% from dermatitis [17]. In another study performed in the US between 1996 and 1997, one in five ambulatory care visits for patients aged more than 55 years resulted in a dermatological diagnosis [18]. The most common single diagnosis in men and women was actinic keratosis, followed by asteatotic dermatitis and nonmelanoma skin cancer. These three diagnoses accounted for 25% of all dermatological diagnoses. Of course, the pattern of skin diseases in the elderly differs among the regions of the world. In developing countries parasite infections are still of major concern [1], while the differences in dermatological diagnoses between industrialized regions in Asia and Canada [14] (Table 1.1) are probably related to a combination of life-style, complexion and other genetic factors.

A major concern is skin cancer, which has increased in incidence in recent decades. This is especially true for cutaneous melanoma. For example, in the early 1970s the Saarland Cancer Registry in Germany reported 3 cases per 100,000 inhabitants per year, but by 1990 reported 9 cases. In Australia (40–60 cases per 100,000 inhabitants per year) and the US (10–20 cases per 100,000 inhabitant per year) the incidences are even higher. In contrast, age distribution has not changed significantly during the last decade and most melanomas are diagnosed at between 50 and 60 years of age [10]. The mean age of patients with nonmelanoma skin cancer is higher than that of patient with melanoma, although data from Florida (US) indicate that the average age at first skin cancer diagnosis decreased from 76.2 years in 1972 to 64.1 years in 2001 [5]. The increasing risk of cutaneous malignancy with age is probably attributable in large part to increased life-time sun exposure. However, a combination of age-associated factors such as decreased immunological surveillance, increased free radicals and loss of DNA repair capacity are also important [11].

1.5 The Challenge Faced by Dermatologists

- Demographic change leads to more skin diseases which have to be treated.
- Increase in cosmetic procedures is not a direct consequence of demographic changes.

An aging population is a challenge also faced by dermatologists. This challenge is intensified by factors other than aging alone. Life-styles have changed during recent decades. Dermatology in the new millennium will also face a change

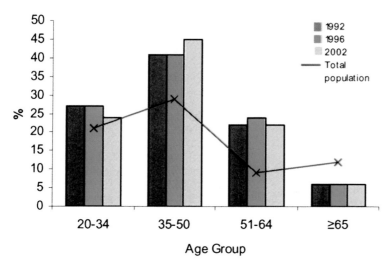

Fig. 1.4. Age distribution of rejuvenation procedures in the US in comparison to the age distribution of the US population (1992–2002)

in terms of racial diversity which is a result of globalization and migration. This is an important aspect considering the effects of solar UV radiation on human skin. Actinic skin damage primarily affects population groups with a Caucasian skin type and this has an impact on the occurrence of actinic keratoses and nonmelanoma skin cancer in Caucasians living in tropical or subtropical regions. The incidence of skin cancers remains very low in darker pigmented populations of African or Asian origin.

It is clear that there are many underrated dermatological needs in the elderly, some of which are of substantial medical importance. However, all of them affect the quality of life of individuals.

Additionally, dermatologists worldwide face a change in patients and their needs. An increasing public awareness about the importance of skin and its appearance has been the result of campaigns in the media. Many elderly have cosmetic concerns such as wrinkling of the skin, furrows, solar lentigines and benign neoplasms. In particular, women visit dermatologists for treatment and prevention of skin aging. The desire for everlasting beauty plays an important role in our societies, and the dermatologist is faced with questions regarding treatment and prevention of skin aging [13].

So-called rejuvenation procedures are more numerous and more often applied. Statistics of the American Society of Plastic Surgery [2]

show that the most popular nonsurgical procedure is Botox injections with more than 2.8 million individuals receiving this treatment in 2003. There was a 1500% increase in cosmetic rejuvenation procedures in the US between 1992 (0.4 million) and 2002 (6.6 million). Botox injections have only been included in this statistic since 2000, but the increase is still 1300% when leaving out these injections. This increase is by no means reflected by the 13% increase in the US population during the same time. There is also no correlation between the age distribution of the US population and the age distribution of people applying for rejuvenation procedures, as Fig. 1.4 demonstrates. Compared to their proportion in the population, people aged 65 years and over had these procedures least often. This is remarkable since age-related skin changes predominantly occur in this age group. The increase in these procedures therefore is not a direct consequence of recent demographic changes and probably is instead a consequence of their acceptance by the population.

1.6 Conclusion

The population is aging at a rapid rate, and with the baby boomer generation at the geriatric doorstep the trend is set to continue in the next 50 years. The elderly are affected by many dermatological concerns, which are not only

1

caused by the normal aging process. Moreover, these concerns are related to additional life-long exposure to environmental and chemical agents. As people advance in life the risk of getting skin diseases increases. Not only is there an increased awareness of skin diseases in our societies, but also growing concern about our appearance in an age influenced by the media. The increasing proportion of men and women interested in rejuvenation is a major challenge for dermatology. The demographic change will have an impact on health economic and political agendas world wide. The need to develop strategies on how to accommodate the aging society is essential and involves a great range of preventive measures from all medical services. An important role for dermatologists is evolving in the prevention of skin diseases and skin cancers in the aging population.

Acknowledgements

Statistics and descriptions of population dynamics are freely available on the internet. We mainly used information from the Federal Institute for Population Research [8], United Nations Population Division [20] and the US Census Bureau [21].

References

1. Abdel-Hafez K, Abdel-Aty MA, Hofny ER (2003) Prevalence of skin diseases in rural areas of Assiut Governorate, Upper Egypt. Int J Dermatol 42:887–892
2. American Society of Plastic Surgery (2004) Age distribution by cosmetic procedure. www.platicsurgery.org
3. Beauregard S, Gilchrest BA (1987) A survey of skin problems and skin care regimes in the elderly. Arch Dermatol 123:1638–1643
4. Berneburg M, Plettenberg H, Krutmann J (2000) Photoaging of human skin. Photodermatol Photoimmunol Photomed 16:239–244
5. Collins GL, Nickoonahand N, Morgan MB (2004) Changing demographics and pathology of nonmelanoma skin cancer in the last 30 years. Semin Cutan Med Surg 23:80–83
6. Centers for Disease Control and Prevention (2003) Public health and aging: trends in aging–United States and worldwide. JAMA 289:1371–1373
7. Dewberry C, Norman RA (2004) Skin cancer in elderly patients. Dermatol Clin 22:93–96
8. Federal Institute for Population Research (2004) Bevölkerung: Fakten–Trends- Ursachen- Erwartungen. Die wichtigsten Fragen. Sonderheft der Schriftenreihe des BiB, 1-52. BiB Bundesinstitut für Bevölkerungsforschung.
9. Fisher GJ, Kang S, Varani J, et al (2002) Mechanisms of photoaging and chronological skin aging. Arch Dermatol 138:1462–1470
10. Garbe C, Blum A (2001) Epidemiology of cutaneous melanoma in Germany and worldwide. Skin Pharmacol Appl Skin Physiol 14:280–290
11. Graham-Brown RA (2004) Diagnosing skin disease in the elderly. Practitioner 974–981
12. Johnson MLT (1986) Skin conditions and related need for medical care among persons 1–74 years. Vital Health Statistics vol. 11, 79-1660 United States, DHEW Publication
13. Kosmadaki MG, Gilchrest BA (2002) The demographics of aging in the United States: implications for dermatology. Arch Dermatol 138:1427–1428
14. Liao YH, Chen KH, Tseng MP, et al (2001) Pattern of skin diseases in a geriatric patient group in Taiwan: a 7-year survey from the outpatient clinic of a university medical center. Dermatology 203:308–313
15. Schäfer T, Merkl J, Klemm E, et al (2002) Prävalenz und Altersabhängigkeit von chronischen und pigmentierten Veränderungen der Haut bei Erwachsenen. Inform Biom Epidemiol Med Biol 33:192
16. Schneider BJ, Norman RA (2004) Cutaneous manifestations of endocrine-metabolic disease and nutritional deficiency in the elderly. Dermatol Clin 22:23–31
17. Smith DR, Atkinson R, Tang S, et al (2002) A survey of skin disease among patients in an Australian nursing home. J Epidemiol 12:336–340
18. Smith ES, Fleischer AB Jr, Feldman SR (2001) Demographics of aging and skin disease. Clin Geriatr Med 17:631–641
19. Theodosat A (2004) Skin diseases of the lower extremities in the elderly. Dermatol Clin 22:13–21
20. United Nations Division for Social Policy and Development (1998) The ageing of the world's population–demographics. http://www.un.org/esa/socdev/ageing/agewpop.htm
21. US Census (2000) Profile of general demographic characteristics: 2000. US Census Bureau. http://www.census.org

Chapter 2

Clinical and Histological Features of Intrinsic versus Extrinsic Skin Aging

2

Mina Yaar

Contents

2.1 Introduction

Aging, a process that results in cellular attrition and senescence eventually terminating in decreased viability and death, is affected by a genetic program as well as by cumulative environmental and endogenous insults that take place throughout the organism's life-span. Understanding the aging process is important, in part because the proportion of individuals 55 years or older is continuously increasing, predicted to be 31% in the USA in the year 2030 [1] with similar demographic shifts predicted for Europe and Japan. Although skin aging comprises only a portion of the entire aging process, data from the National Ambulatory Medical Care Survey in the USA show that during the years 1996–1997, 4.6% of total physician visits were related to dermatological problems (reviewed in reference 2) and this percentage is anticipated to be even higher as the proportion of elderly in the population increases, leading to a significant burden on dermatologists and primary care physicians.

As life expectancy increases, forcing older individuals to postpone their retirement and/or plan for a lengthy retirement, the elderly seek intervention modalities to improve their appearance and reverse the signs of aging. As a result, a substantial increase in the number of visits to beauticians, dermatologists and plastic surgeons is anticipated in the future. To effectively manage skin diseases of the elderly and to use the proper intervention modalities to reverse cutaneous aging, it is important to be familiar with the clinical and histological changes that accompany skin aging.

2

2.2 Chronological Skin Aging: Definition and Pathophysiology

Chronological skin aging comprises those changes in the skin that occur as a result of passage of time alone. These changes occur in part as the result of cumulative endogenous damage from continuous formation of reactive oxygen species (ROS) generated during oxidative cellular metabolism. Despite intricate cellular antioxidant defense systems, generated ROS damage several cellular constituents including membranes, enzymes and DNA, and also interfere with DNA–protein and protein–protein interactions (reviewed in reference 3).

Telomeres, the terminal portions of eukaryotic chromosomes, have been implicated in changes that occur as a result of chronological aging. With each cell division human telomere length shortens. Even in relatively quiescent skin fibroblasts more than 30% of the telomere length is lost during adulthood [4]. Critically short telomeres signal cell cycle arrest or apoptosis, depending on cell type, contributing to cellular depletion with aging [3].

In a similar manner to aging of other systems, chronological skin aging is affected by modifications in growth factors and hormones that decline with age. The best recorded decline is that of sex steroids such as estrogen, testosterone, dehydroepiandrosterone (DHEA) and its sulfate ester DHEAS [5–7]. However, other hormones including melatonin, cortisol, thyroxine [5], growth hormone [7] and insulin like growth factor I [8, 9] decline as well. Also, the active form of vitamin D, 1,25-dihydroxyvitamin D_3, a molecule that affects a variety of tissues including the skin in a way distinct from its effect on calcium homeostasis (see below), declines with age [7]. In addition to the decline in their constitutive levels, induced levels of certain signaling molecules including cytokines and chemokines decline with age leading to deterioration of several skin functions (see below) [10].

Apart from the age-associated decline in synthesis and secretion of signaling molecules, the levels of their receptors decline as well. The levels of receptors for vitamin D [11], beta adrenergic compounds [12], neurotransmitters [13] and dopamine [14] decrease with age. In the epidermis, there is a decline in the expression and synthesis of receptors for the interleukin-1 cytokine family [15]. Also, in vitro late-passage fibroblasts display decreased levels of receptors for low-density lipoprotein [16] and platelet-derived growth factor [17], and fibroblasts derived from elderly individuals have decreased levels of epidermal growth factor receptors [18].

However, decline in signaling molecules and their receptors is by no means the general rule for changes that occur with aging, as there are signaling molecules that increase with age. Examples include transforming growth factor-beta1, a cytokine that has been implicated in inducing fibroblast senescence [19], and caveolin, a membrane-associated scaffolding protein (reviewed in reference 20). Indeed, cellular attrition and senescence that characterize the aging process are the result of molecular alterations in the cellular milieu as well as in DNA and proteins within the cell. These changes lead to aberrant cellular responses to environmental changes, eventually resulting in decreased viability and death.

2.3 Clinical and Histological Manifestations of Chronologically Aged Skin

Clinical manifestations of chronologically aged skin include xerosis, laxity, wrinkles, slackness and the appearance of a variety of benign neoplasms such as seborrheic keratoses and cherry angiomas.

The histological features that accompany these clinical manifestations are summarized in Table 2.1. There are no obvious alterations in stratum corneum structure [21] and relatively little change in epidermal thickness, keratinocyte shape and corneocyte cohesion [21]. However, there is substantial loss of melanocytes and Langerhans cells [3].

The major cutaneous changes in chronologically aged skin are observed in the dermoepidermal junction which displays flattening of the rete ridges leading to reduced surface contact between the epidermis and dermis and, as a result, reduced exchange of nutrients and metabolites between these compartments.

Table 2.1. Histological features of aging human skin

Epidermis	Dermis	Appendages
Flattened dermoepidermal junction	Atrophy (loss of dermal volume)	Depigmented hair
Variable thickness	Alteration of connective tissue structure	Loss of hair
Variable cell size and shape	Fewer fibroblasts	Conversion of terminal to vellus hair
Occasional nuclear atypia	Fewer mast cells	Abnormal nailplates
Fewer melanocytes	Fewer blood vessels	Fewer glands
Fewer Langerhans cells	Shortened capillary loops	
	Abnormal nerve endings	

Modified with permission of McGraw-Hill, Inc. (for original publication, see reference 3)

The dermis appears hypocellular with fewer fibroblasts and mast cells and loss of dermal volume. Electron microscopy studies have revealed that collagen fibers become loose and there is a moderate increase and thickening of elastic fibers with resorption of most subepidermal fibers [3, 21]. In addition, there is a decrease in the number of dermal blood vessels, a shortening of capillary loops, and a decrease in the density of Pacinian and Meissner's corpuscles, the cutaneous end organs responsible for pressure and light touch perception [3]. Finally, there is loss of sensory and autonomic innervation involving both the epidermis and dermis (reviewed in reference 22).

Modifications in cutaneous appendages include hair loss reflecting conversion of terminal hair to vellus hair [3]. Also, there is graying of the hair as a result of melanocyte loss from the hair bulb [3] and aberrant melanocyte function including decreased tyrosinase activity, reduced and less-efficient melanosomal transfer and faulty melanocyte migration and/or proliferation from a presumptive storage area to an area close to the dermal papilla [23].

2.3.1 Wrinkles

Intrinsic factors that affect facial structures and contribute to facial wrinkle formation include changes in expression muscles, loss of subcutaneous fat, persistent gravitational force and loss of facial bone and cartilage. Expression lines occur as a result of repeated traction exerted by facial muscles that ultimately leads to the formation of deep creases over the forehead and between the eyebrows (Fig. 2.1), periorbitally and in the nasolabial folds. Histologically, thick hypodermal connective tissue strands containing muscle cells are present underneath the wrinkle [24]. In addition, evidence suggests that with aging, changes take place in the musculoaponeurotic structures leading to increased limpness and resulting in exaggeration of certain expression wrinkles such as those of the nasolabial folds. Indeed, like other striated muscles, facial muscles show accumulation of the "age pigment" lipofuscin, a marker of cellular damage, and this muscle deterioration with age compounded by diminished neuromuscular control contributes to wrinkle formation [25].

The force of gravity that continuously acts on the body influences the skin affecting the distribution of facial soft tissues resulting in sagging of the skin. Also, as the skin becomes increasingly lax with age and soft tissue support diminishes, the force of gravity becomes an important factor. Gravity exerts a mechanical force pulling down facial skin leading to the formation of sagging, loose skin (Fig. 2.2). Skin sagging is particularly prominent in the upper and lower eyelids, giving the eyelid a "baggy" appearance, in the cheeks, and in the neck area. However, it should be noted that not all investigators agree that gravity is the only culprit in

2

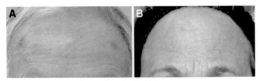

Fig. 2.1. Expression wrinkles. Repeated traction by facial muscles results in the formation of creases over the forehead (**A**) and between the eyebrows (**B**)

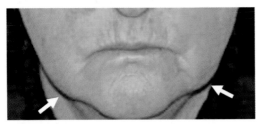

Fig. 2.2. Gravitational wrinkles. The force of gravity together with age-associated cutaneous laxity result in jowls

the formation of sagging skin. It is argued that gravity merely allows us to see the changes that have occurred as a result of fat depletion and redistribution [26].

Indeed, with aging fat is depleted from certain facial areas including the forehead, preorbital, buccal, temporal and perioral regions. In contrast, there is a prominent increase in the bulk of fatty tissue in other areas including the submental regions, the jowls, the nasolabial folds and the lateral malar areas. In contrast to the appearance of a young face where fat is diffusely dispersed, in aging facial skin fat tends to accumulate in pockets, and subsequently when this excess fat is subject to the force of gravity, sagging and drooping of the skin occurs [26].

Finally, like other parts of the skeleton, facial bones display reduced mass with age. Bone resorption particularly affects the mandible, maxilla and frontal bones. Bone loss in these areas enhances facial skin droopiness and contributes to the obliteration of the demarcation between the contour of the jaw and the neck that is so distinct in young adults [27].

The skin of older individuals also displays an array of fine superficial lines that characteristically disappear when the skin is stretched.

Histologically, the epidermis appears atrophic as a result of decreased epidermal turnover rate. There is resorption of the elastic fiber network in the subepidermal area, and the reticular dermis displays atrophic collagen bundles. In the dermis, remaining fibroblasts appear shriveled [3]. Also, there are alterations in connective tissue structure reflecting decreased collagen synthesis and increased levels of metalloproteinases [28]. Interestingly, a correlation has been found between degenerative changes in dermal elastic fiber and menopause, suggesting that estrogen and/or progesterone contribute to elastic fiber maintenance [29].

2.3.2 Benign Age-associated Neoplasms

2.3.2.1 Seborrheic Keratoses

Seborrheic keratoses are benign epithelial neoplasms that are monoclonal in origin [30]. They start as flat hyperpigmented macules and progress to become hyperkeratotic verrucous plaques highly variable in size and color. Seborrheic keratoses first appear in the third to fifth decade of life and become increasingly numerous throughout life, independent of sun exposure (reviewed in reference 31). They are regarded as the best biomarker of intrinsic skin aging. Presumably they represent a focal subtle loss of homeostasis, with resulting over-proliferation of keratinocytes and melanocytes, although the pathogenesis is not known. Recently, keratinocytes with basaloid morphology in seborrheic keratoses have been reported to express high levels of endothelin-1 (ET-1), associated with increased tyrosinase expression in the melanocytes, compared to control perilesional skin [32], suggesting that ET-1-induced melanogenesis [33], dendricity [34], and melanocyte proliferation [35] may play a role in the evolution of this neoplasm.

2.3.2.2 Cherry Angiomas

Cherry angiomas are small red to purple vascular malformations composed of venous capillaries and postcapillary venules that are present in the dermal papillae and are connected to each other and to the venular portion of the superficial vascular plexus [36]. They contain increased numbers of mast cells as compared to surrounding skin [37], and as mast cells are known to synthesize proangiogenic factors such as vascular endothelial growth factor and basic fibroblast growth factor [38, 39], they may be causally involved in the development of these age-associated lesions.

2.4 Photoaging: Definition and Pathophysiology

Photoaging comprises those changes in the skin that are the result of chronic sun exposure superimposed on chronological skin aging [3]. Photoaging occurs as a result of cumulative damage from ultraviolet (UV) radiation.

UV radiation (wavelengths in the 100–400 nm range) comprises only 5% of the terrestrial solar radiation. It is arbitrarily divided into UVA (320–400 nm), UVB (280–320 nm) and UVC (100–280 nm). The UVC portion of the spectrum is not present in terrestrial sunlight, except at high altitudes, as it is absorbed by the atmospheric ozone layer (reviewed in reference 40). The predominant component of solar UV radiation is UVA, the intensity of which varies little with season or time of day, and unlike UVB radiation is not blocked by glass [41]. Although the energy per photon delivered as UVA radiation is approximately 1,000-fold less than that as UVB (reviewed in reference 42), because of its longer wavelengths, UVA penetrates the skin to reach deeper dermal layers [42].

Most adverse effects of UVA in the skin are assumed to be the result of oxidative damage [43–47] mediated through UVA absorption by cellular chromophores such as urocanic acid [48], riboflavin [49] and melanin precursors [50] that act as photosensitizers [44, 45] leading to the generation of reactive oxygen species (ROS) and free radicals (reviewed in reference 51). Interestingly, the end products of advanced glycation that accumulate with aging in long-lived proteins such as those of the extracellular matrix also act as UVA chromophores to be-

come photosensitizers affecting dermal fibroblasts [52–54]. However, although it is evident that UVA plays a role in cutaneous photodamage, it is not clear what changes are the result of UVA irradiation and what changes are induced by UVB irradiation. Studies by Lavker et al. [55] suggest that UVA radiation when administered over time can induce changes similar to those induced by UVB, including epidermal hyperplasia, stratum corneum thickening, Langerhans cell depletion, dermal inflammation and accumulation of lysozymes on dermal fibers.

UVB irradiation primarily affects the epidermis. It is directly absorbed by cellular DNA, leading to the formation of DNA lesions, primarily cyclobutane dimers and pyrimidine (6-4) pyrimidone photoproducts (reviewed in reference 56). Despite comprehensive nuclear DNA damage repair systems, DNA damage is rarely completely repaired. When cells sustain abundant DNA damage they undergo apoptosis (reviewed in reference 57), a process mediated largely by the tumor suppressor p53 protein (reviewed in references 58 and 59). p53 also participates in DNA damage repair and in transient cell cycle arrest after DNA damage [59]. However, those cells that have not undergone apoptosis and in which the damage is not completely repaired risk developing mutations and eventually becoming cancerous. This is particularly important in view of recent epidemiological studies showing that more than 90% of epidermal squamous cell carcinomas and more than 50% of basal cell carcinomas display UV-induced mutations that inactivate p53 (reviewed in reference 60). Furthermore, p53 mutations are present in premalignant actinic keratoses (reviewed in references 61 and 62), suggesting that p53 mutations occur early, increasing the risk of malignant transformation of affected cells.

Apart from its direct effect on epidermal DNA, studies in the murine system have shown that UVB irradiation affects both the cutaneous and systemic immune responses leading to defective antigen presentation and formation of suppressor T-cells, allowing the propagation of cancerous cells that would otherwise be rejected (reviewed in references 63 and 64). In this regard, it has been suggested that UVA, by inducing lipid peroxidation, stimulates the outward

Table 2.2. Features of actinically damaged skin (basal cell carcinoma and squamous cell carcinoma also occur in actinically damaged skin but, unlike the table entries, affect only a small minority of individuals with photoaging)

2

Clinical abnormality	Histological abnormality
Dryness (roughness)	Increased compaction of stratum corneum; increased thickness of granular cell layer; reduced epidermal thickness; reduced epidermal mucin content
Actinic keratoses	Nuclear atypia; loss of orderly, progressive keratinocyte maturation; irregular epidermal hyperplasia and/or hypoplasia; occasional dermal inflammation
Irregular pigmentation	
Freckling	Increased number of hypertrophic, strongly dopa-positive melanocytes
Lentigines	Elongation of epidermal rete ridges; increases in number and melanization of melanocytes
Guttate hypomelanosis	Reduced number of atypical melanocytes
Persistent hyperpigmentation	Increased number of dopa-positive melanocytes and increased melanin content per unit area; increased number of dermal melanophages
Wrinkling	Decreased and degraded collagen; increased matrix-degrading metalloproteinases; contraction of septae in the subcutaneous fat
Stellate pseudoscars	Absence of epidermal pigmentation; altered fragmented dermal collagen
Fine nodularity and/or coarseness	Nodular aggregations of fibrous to amorphous elastotic material in the papillary dermis
Inelasticity	Elastotic dermis
Telangiectasia	Ectatic vessels often with atrophic walls
Venous lakes	Ectatic vessels often with atrophic walls
Purpura (easy bruising)	Extravasated erythrocytes and increased perivascular inflammation
Comedones (maladie de Favre et Racouchot)	Ectatic superficial portion of the pilosebaceous follicle
Sebaceous hyperplasia	Concentric hyperplasia of sebaceous glands

Modified with permission from McGraw-Hill, Inc. (for original publication, see reference 3)

migration of immune responsive cells from the epidermis and thus further contributes to immunosuppression [63]. Also, UVB irradiation induces the secretion of epidermal cytokines, and evidence suggests that of the induced cytokines, tumor necrosis factor-α and interleukin-10 play a major role in UVB-induced immunosuppression (reviewed in reference 63).

With regard to mechanisms that mediate changes observed in normal photodamaged skin, in addition to the above, UV irradiation activates cell surface receptors. This leads to propagation of intracellular signaling and the synthesis of transcription factors, nuclear proteins that bind the DNA to enhance or repress gene transcription (reviewed in reference 65). One transcription factor that is quickly and prominently induced by UV irradiation is AP-1. AP-1 interferes with collagen gene transcription in fibroblasts, decreasing the levels of the major procollagens I and III. In addition, AP-1 stimulates the transcription of genes that encode matrix-degrading enzymes such as metalloproteinases [65].

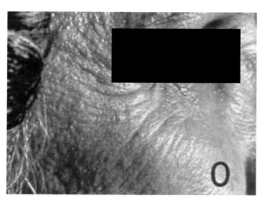

Fig. 2.3. Elastotic wrinkles. Coarse tanned skin displays deep creases and fine nodularity

2.5 Clinical and Histological Manifestations of Photoaged Skin

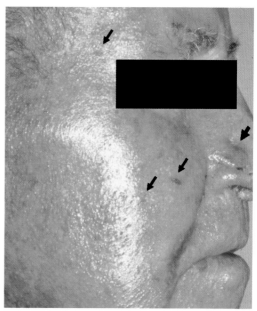

Fig. 2.4. Atrophic response to photodamage. An individual with skin type I displays smooth atrophic skin devoid of wrinkles. There are multiple actinic keratoses (*thin arrows*) and a scar (*thick arrow*) in an area previously involved with basal cell carcinoma

Because skin areas that are exposed to the sun are also those that are visually apparent, the perception of an individual's age is primarily influenced by the amount of his/her cutaneous photodamage. The clinical and histological features of photodamaged skin are summarized in Table 2.2. Interestingly, the response to UV-induced damage appears to depend on the individual's skin type. Individuals with skin types III–V show "hyperplastic" responses – they display thick, leathery skin with coarse wrinkles. Sometimes there is also fine nodularity (elastosis) and comedones (*maladie de-Favre et Racouchot*). The skin appears permanently hyperpigmented or "bronzed" with a yellow to reddish hue and it displays multiple hyperpigmented macules (lentigines). Elastotic wrinkles characterize photodamaged skin of individuals with skin types III–V. Clinically, they form a crisscross rhomboidal pattern and the coarse skin displays fine nodularity (Fig. 2.3). Histologically, there is irregular epidermal thickness. The papillary dermis displays nodular aggregations of fibrous to amorphous abnormal elastotic material [3]. Also, the amount of glycosaminoglycans and proteoglycans in the dermal ground substance is increased while collagen fibers decrease and become partially degraded as a result of UV-induced synthesis

and secretion of matrix-degrading metalloproteinases [3].

Individuals with skin types I–II show an "atrophic," "exhaustive," "dysplastic" response – they display fewer wrinkles and have a smoother skin but with premalignant lesions primarily actinic keratoses, as well as epidermal malignancies (Fig. 2.4) (reviewed in references 3 and 66). Although the molecular basis of the different photoaging patterns in fair versus dark skin types is unknown, it has been hypothesized that they reflect different degrees of inducible DNA repair capacity following UV exposure [67].

One of the most prominent histological features of photodamage is solar elastosis (Fig. 2.5). Elastosis is a material composed of large tangled masses of degraded elastic tissue. In addition, there are increased amounts of ground substance that is primarily made of glycosaminoglycans and proteoglycans. Interestingly, immediately below the epidermis, in the uppermost portion of the dermis, there is a narrow band containing an eosinophilic substance that is

2

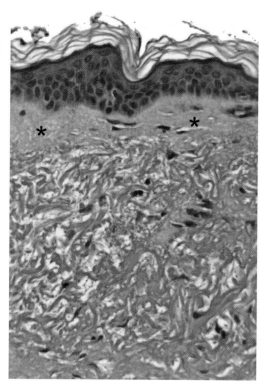

Fig. 2.5. Photodamaged skin. H&E staining of photodamaged skin displaying bluish masses of deranged elastic fibers characteristic of solar elastosis. A thin subepidermal Grenz zone (*asterisks*) is present. Courtesy of Jag Bhawan, MD. Reproduced with permission of McGraw-Hill, Inc. (for original publication, see reference 3)

composed primarily of glycosaminoglycans and newly formed collagen and is called Grenz zone [3, 65]. It is considered to represent an area where active repair of photodamage takes place and histologically is reminiscent of scar tissue in wounds. Deeper in the dermis collagen fibers appear degraded, clumped and fragmented [68]. Also, the dermis frequently displays abundant inflammatory infiltrate composed of mast cells, histiocytes and other mononuclear cells [3].

2.5.1 Premalignant Neoplasms: Actinic Keratoses

Actinic keratoses are epidermal neoplasms that display proliferation of cytologically abnormal keratinocytes. Clinically, they appear as erythematous macules and papules with adherent coarse scale on a background of photodamaged skin. The abnormal keratinocytes in actinic keratoses are the result of UV-induced mutations in the tumor suppressor gene *p53* [61]. The protein encoded by *p53* is important for cell cycle arrest after DNA damage; it also plays as major role in apoptosis induction of severely damaged cells (reviewed in reference 59). The presence of *p53* mutations allows affected cells to proliferate despite having sustained DNA damage, risking the formation of additional mutations and eventually evolving into squamous cell carcinoma.

2.6 Age-Associated Decline in Skin Functions

2.6.1 Cell Replacement and Wound Healing

Keratinocytes comprise 90% of the epidermal cell population. With time, they lose their proliferative capacity [3], the ability to properly terminally differentiate in order to form the protective stratum corneum [3], and the ability to elaborate cytokines and other cell–cell signals in response to environmental stimuli (reviewed in reference 69). Presumably, this deterioration contributes to the slow healing of minor injuries and weaker surgical scars, as well as a tendency to non-healing ulcers. Compromised wound healing also appears to be affected by decreased function of macrophages and T-cells whose ability to penetrate into the wound bed is compromised. This is compounded by decreased chemokine production [10] and decreased neurogenic inflammation accompanied by decreased neuropeptide synthesis and secretion – all important for appropriate tissue repair [70].

2.6.2 Sensory Function

With aging, there is a decrease in sensory perception for light touch, vibratory sensation, the ability to discriminate two points and spatial acuity, and an increase in pain threshold [3, 71].

Although the exact mechanism that underlies these changes is not yet completely understood, studies have shown that in people aged 60 years or older there is decreased density of both unmyelinated and myelinated nerve fibers that transmit thermal and noxious sensations [72]. In addition, there is a decrease in the synthesis and transport of neuropeptides such as substance P and calcitonin gene-related peptide. Overall, differences in sensory perception between young and old individuals appear to be greatest when the stimuli are of short duration and when they involve the extremities.

2.6.3 DNA Damage Repair

It is well documented that DNA damage and mutation frequencies increase with age. Although mutation accumulation could be the result of passage of time alone, data suggest that the capacity to repair damaged DNA decreases with age. DNA repair capacity has been found to decrease 0.61% per year in peripheral blood lymphocytes, and people who develop basal cell carcinomas (BCC) at an earlier age have decreased repair capacity as compared with individuals who develop BCC later in life [73]. This could be the result of an age-associated decrease in the level of proteins that participate in nucleotide excision repair, as has been reported for aged dermal fibroblasts [74]. Together the studies point to deficient DNA repair capacity predisposing the elderly to the development of cancer.

2.6.4 Immune Function

With aging there is a reduction in the number of epidermal Langerhans cells, the skin's immune antigen-presenting effector cells [3]. There is also decreased production of the epidermal cytokine interleukin (IL)-1α and as a result decreased production of downstream cytokines including IL-6, granulocyte-macrophage colony stimulating factor and IL-8, among others (reviewed in reference 75). Evidence also suggests that with aging, both the cell-mediated and humoral immunity deteriorate, with T-cells displaying re-

duced proliferative capacity and cytokine production in response to triggers [76], and failure to appropriately select antigen-activated B-cells in the germinal centers of the lymph nodes [77]. These decrements compromise the immune system of the elderly, rendering them more susceptible to infections [78] and as a result of decreased immune surveillance possibly also more susceptible to the development of cancer.

2.6.5 Vitamin D Production

The human epidermis plays a role in the generation of the active form of vitamin D, $1,25(OH)_2D_3$ [3]. Besides its role in calcium homeostasis and bone maintenance $1,25(OH)_2D_3$ has been implicated in immune responses, affecting macrophage function and modulating the release of inflammatory cytokines [79, 80], and possibly in prevention of certain cancers of epithelial origin such as breast and colon [81, 82]. In this context it is important to note that elderly individuals have reduced serum levels of vitamin D [83], in part because of decreased consumption of vitamin D in their diet, and in part because of insufficient sun exposure. Furthermore, the level of the epidermal precursor of vitamin D, 7-dehydrocholesterol per unit skin surface decreases linearly by approximately 75% between early and late adulthood [84], suggesting that due to lack of precursor, older individuals may fail to synthesize sufficient amounts of $1,25(OH)_2D_3$. Vitamin D and calcium supplementation are thus particularly important in this segment of the population.

2.6.6 Barrier Function and Mechanical Protection

Under baseline conditions barrier function is conserved even in aged skin. Age-associated differences are more obvious after injury as the time required to reconstitute an effective barrier following trauma such as tape stripping increases markedly, from approximately 3 days in young adults to more than 6 days on average in elderly adults [85]. Biochemical and ultrastructural studies have revealed that the decreased

2

rate of barrier regeneration is due at least in part to decreased lipid synthetic capacity with aging [85]. A significant influence of female hormones on stratum corneum sphingolipid composition has also been reported [86], with a presumed impact on postmenopausal skin.

Compromised thermoregulation predisposes the elderly to life-threatening conditions including heat stroke and hypothermia. Decreased sweat production with age additionally predisposes the elderly to heat stroke. Finally, due to decreased gonadal and adrenal androgens, sebum production decreases approximately 23% per decade beginning in the second decade – a 60% decrease over an adult life-span [3].

References

1. US Census Bureau, online database. www.census.gov

2. Smith ES, Fleischer AB Jr, Feldman SR (2001) Demographics of aging and skin disease. Clin Geriatr Med 17:631–641

3. Yaar M, Gilchrest BA (2003) Aging of skin. In: Freedberg IM, Eisen AZ, Wolff K, Austen KF, Goldsmith LA, Katz SI (eds) Fitzpatrick's dermatology in general medicine, vol 2. McGraw-Hill, New York, pp 1386–1398

4. Allsopp RC, Vaziri H, Patterson C, Goldstein S, Younglai EV, Futcher AB, Greider CW, Harley CB (1992) Telomere length predicts replicative capacity of human fibroblasts. Proc Natl Acad Sci U S A 89:10114–10118

5. Wespes E, Schulman CC (2002) Male andropause: myth, reality, and treatment. Int J Impot Res 14 Suppl 1:S93–98

6. Phillips TJ, Demircay Z, Sahu M (2001) Hormonal effects on skin aging. Clin Geriatr Med 17:661–672

7. Arlt W, Hewison M (2004) Hormones and immune function: implications of aging. Aging Cell 3:209–216

8. Arvat E, Broglio F, Ghigo E (2000) Insulin-like growth factor I: implications in aging. Drugs Aging 16:29–40

9. Khan AS, Sane DC, Wannenburg T, Sonntag WE (2002) Growth hormone, insulin-like growth factor-1 and the aging cardiovascular system. Cardiovasc Res 54:25–35

10. Swift ME, Burns AL, Gray KL, DiPietro LA (2001) Age-related alterations in the inflammatory response to dermal injury. J Invest Dermatol 117:1027–1035

11. Bell MN, Jackson AG (1995) Role of vitamin D in the pathogenesis and treatment of osteoporosis. Endocr Pract 1:44–47

12. Schutzer WE, Mader SL (2003) Age-related changes in vascular adrenergic signaling: clinical and mechanistic implications. Ageing Res Rev 2:169–190

13. Peters A (2002) Structural changes that occur during normal aging of primate cerebral hemispheres. Neurosci Biobehav Rev 26:733–741

14. Ingram DK (2000) Age-related decline in physical activity: generalization to nonhumans. Med Sci Sports Exerc 32:1623–1629

15. Barland CO, Zettersten E, Brown BS, Ye J, Elias PM, Ghadially R (2004) Imiquimod-induced interleukin-1 alpha stimulation improves barrier homeostasis in aged murine epidermis. J Invest Dermatol 122:330–336

16. Bose C, Bhuvaneswaran C, Udupa KB (2004) Altered mitogen-activated protein kinase signal transduction in human skin fibroblasts during in vitro aging: differential expression of low-density lipoprotein receptor. J Gerontol A Biol Sci Med Sci 59:126–135

17. Yeo EJ, Jang IS, Lim HK, Ha KS, Park SC (2002) Agonist-specific differential changes of cellular signal transduction pathways in senescent human diploid fibroblasts. Exp Gerontol 37:871–883

18. Reenstra WR, Yaar M, Gilchrest BA (1993) Effect of donor age on epidermal growth factor processing in man. Exp Cell Res 209 118–122

19. Frippiat C, Chen QM, Zdanov S, Magalhaes JP, Remacle J, Toussaint O (2001) Subcytotoxic H2O2 stress triggers a release of transforming growth factor-beta 1, which induces biomarkers of cellular senescence of human diploid fibroblasts. J Biol Chem 276:2531–2537

20. Park WY, Cho KA, Park JS, Kim DI, Park SC (2001) Attenuation of EGF signaling in senescent cells by caveolin. Ann N Y Acad Sci 928:79–84

21. Lavker RM, Zheng PS, Dong G (1987) Aged skin: a study by light, transmission electron, and scanning electron microscopy. J Invest Dermatol 88:44s–51s

22. Ulfhak B, Bergman E, Fundin BT (2002) Impairment of peripheral sensory innervation in senescence. Auton Neurosci 96:43–49

23. Tobin DJ, Paus R (2001) Graying: gerontobiology of the hair follicle pigmentary unit. Exp Gerontol 36:29–54

24. Pierard GE, Uhoda I, Pierard-Franchimont C (2003) From skin microrelief to wrinkles. An area ripe for investigation. J Cosmet Dermatol 2:21–28

25. Dayan D, Abrahami I, Buchner A, Gorsky M, Chimovitz N (1988) Lipid pigment (lipofuscin) in human perioral muscles with aging. Exp Gerontol 23:97–102

26. Donofrio LM (2000) Fat distribution: a morphologic study of the aging face. Dermatol Surg 26:1107–1112

27. Ramirez OM, Robertson KM (2001) Comprehensive approach to rejuvenation of the neck. Facial Plast Surg 17:129–140

28. Varani J, Warner RL, Gharaee-Kermani M, Phan SH, Kang S, Chung JH, Wang ZQ, Datta SC, Fisher GJ, Voorhees JJ (2000) Vitamin A antagonizes decreased cell growth and elevated collagen-degrading matrix metalloproteinases and stimulates collagen accumulation in naturally aged human skin. J Invest Dermatol 114:480–486

29. Bolognia JL, Braverman IM, Rousseau ME, Sarrel PM (1989) Skin changes in menopause. Maturitas 11:295–304

30. Nakamura H, Hirota S, Adachi S, Ozaki K, Asada H, Kitamura Y (2001) Clonal nature of seborrheic keratosis demonstrated by using the polymorphism of the human androgen receptor locus as a marker. J Invest Dermatol 116:506–510

31. Silver SG, Ho VCY (2003) Benign epithelial tumors. In: Freedberg IM, Eisen AZ, Wolff K, Austen KF, Goldsmith LA, Katz SI (eds) Fitzpatrick's dermatology in general medicine, vol 1. McGraw-Hill, New York, pp 767–785

32. Teraki E, Tajima S, Manaka I, Kawashima M, Miyagishi M, Imokawa G (1996) Role of endothelin-1 in hyperpigmentation in seborrhoeic keratosis. Br J Dermatol 135:918–923

33. Imokawa G, Miyagishi M, Yada Y (1995) Endothelin-1 as a new melanogen: coordinated expression of its gene and the tyrosinase gene in UVB-exposed human epidermis. J Invest Dermatol 105:32–37

34. Hara M, Yaar M, Gilchrest BA (1995) Endothelin-1 of keratinocyte origin is a mediator of melanocyte dendricity. J Invest Dermatol 105:744–748

35. Imokawa G, Yada Y, Miyagishi M (1992) Endothelins secreted from human keratinocytes are intrinsic mitogens for human melanocytes. J Biol Chem 267:24675–24680

36. Braverman IM, Ken-Yen A (1983) Ultrastructure and three-dimensional reconstruction of several macular and papular telangiectases. J Invest Dermatol 81:489–497

37. Hagiwara K, Khaskhely NM, Uezato H, Nonaka S (1999) Mast cell "densities" in vascular proliferations: a preliminary study of pyogenic granuloma, portwine stain, cavernous hemangioma, cherry angioma, Kaposi's sarcoma, and malignant hemangioendothelioma. J Dermatol 26:577–586

38. Esposito I, Menicagli M, Funel N, Bergmann F, Boggi U, Mosca F, Bevilacqua G, Campani D (2004) Inflammatory cells contribute to the generation of an angiogenic phenotype in pancreatic ductal adenocarcinoma. J Clin Pathol 57:630–636

39. Furuta S, Vadiveloo P, Romeo-Meeuw R, Morrison W, Stewart A, Mitchell G (2004) Early inducible nitric oxide synthase 2 (NOS 2) activity enhances ischaemic skin flap survival. Angiogenesis 7:33–43

40. Diffey BL (2002) Sources and measurement of ultraviolet radiation. Methods 28:4–13

41. Midelfart K, Moseng D, Kavli G, Volden G (1984) One-year measurements of solar UVB and UVA radiation at latitude 70 degrees north. Photodermatology 1:252–254

42. Kaminer MS (1995) Photodamage: magnitude of the problem. In: Gilchrest BA (ed) Photodamage. Blackwell, Cambridge, MA, pp 1–11

43. Tyrrell RM (1995) Ultraviolet radiation and free radical damage to skin. Biochem Soc Symp 61:47–53

44. Kvam E, Tyrrell RM (1997) Induction of oxidative DNA base damage in human skin cells by UV and near visible radiation. Carcinogenesis 18:2379–2384

45. Scharffetter-Kochanek K, Wlaschek M, Brenneisen P, Schauen M, Blaudschun R, Wenk J (1997) UV-induced reactive oxygen species in photocarcinogenesis and photoaging. Biol Chem 378:1247–1257

46. de Gruijl FR (2000) Photocarcinogenesis: UVA vs UVB. Methods Enzymol 319:359–366

47. Agar NS, Halliday GM, Barnetson RS, Ananthaswamy HN, Wheeler M, Jones AM (2004) The basal layer in human squamous tumors harbors more UVA than UVB fingerprint mutations: a role for UVA in human skin carcinogenesis. Proc Natl Acad Sci U S A 101:4954–4959

48. Menon EL, Morrison H (2002) Formation of singlet oxygen by urocanic acid by UVA irradiation and some consequences thereof. Photochem Photobiol 75:565–569

49. Sato K, Taguchi H, Maeda T, Minami H, Asada Y, Watanabe Y, Yoshikawa K (1995) The primary cytotoxicity in ultraviolet-a-irradiated riboflavin solution is derived from hydrogen peroxide. J Invest Dermatol 105:608–612

50. Kipp C, Young AR (1999) The soluble eumelanin precursor 5,6-dihydroxyindole-2-carboxylic acid enhances oxidative damage in human keratinocyte DNA after UVA irradiation. Photochem Photobiol 70:191–198

51. Gasparro FP (2000) Sunscreens, skin photobiology, skin cancer: the need for UVA protection and evaluation of efficacy. Environ Health Perspect 108 [Suppl 1]:71–78

52. Wondrak GT, Roberts MJ, Jacobson MK, Jacobson EL (2002) Photosensitized growth inhibition of cultured human skin cells: mechanism and suppression of oxidative stress from solar irradiation of glycated proteins. J Invest Dermatol 119:489–498

53. Wondrak GT, Roberts MJ, Cervantes-Laurean D, Jacobson MK, Jacobson EL (2003) Proteins of the extracellular matrix are sensitizers of photo-oxidative stress in human skin cells. J Invest Dermatol 121:578–586

54. Wondrak GT, Roberts MJ, Jacobson MK, Jacobson EL (2004) 3-Hydroxypyridine chromophores are endogenous sensitizers of photooxidative stress in human skin cells. J Biol Chem 279:30009–30020

55. Lavker RM, Gerberick GF, Veres D, Irwin CJ, Kaidbey KH (1995) Cumulative effects from repeated exposures to suberythemal doses of UVB and UVA in human skin. J Am Acad Dermatol 32:53–62

56. Eller MS (1995) Repair of DNA photodamage in human skin. In: Gilchrest BA (ed) Photodamage. Blackwell, Cambridge, MA, pp 26–50

57. Gilchrest BA, Eller MS, Geller AC, Yaar M (1999) The pathogenesis of melanoma induced by ultraviolet radiation. N Engl J Med 340:1341–1348

58. Kulms D, Schwarz T (2000) Molecular mechanisms of UV-induced apoptosis. Photodermatol Photoimmunol Photomed 16:195–201

59. Hofseth LJ, Hussain SP, Harris CC (2004) p53: 25 years after its discovery. Trends Pharmacol Sci 25:177–181

60. Brash DE, Ziegler A, Jonason AS, Simon JA, Kunala S, Leffell DJ (1996) Sunlight and sunburn in human skin cancer: p53, apoptosis, and tumor promotion. J Investig Dermatol Symp Proc 1:136–142

61. Ortonne JP (2002) From actinic keratosis to squamous cell carcinoma. Br J Dermatol 146 [Suppl 61]:20–23

62. Ziegler A, Jonason AS, Leffell DJ, Simon JA, Sharma HW, Kimmelman J, Remington L, Jacks T, Brash DE (1994) Sunburn and p53 in the onset of skin cancer. Nature 372:773–776

63. Granstein RD (2003) Photoimmunology. In: Freedberg IM, Eisen AZ, Wolff K, Austen KF, Goldsmith LA, Katz SI (eds) Fitzpatrick's dermatology in general medicine, vol 1. McGraw-Hill, New York, pp 378–386

64. Bergstresser PR (1995) Immediate and delayed effects of UVR on immune responses in skin. In: Gilchrest BA (ed) Photodamage. Cambridge, MA, pp 81–99

65. Fisher GJ, Kang S, Varani J, Bata-Csorgo Z, Wan Y, Datta S, Voorhees JJ (2002) Mechanisms of photoaging and chronological skin aging. Arch Dermatol 138:1462–1470

66. Brooke RC, Newbold SA, Telfer NR, Griffiths CE (2001) Discordance between facial wrinkling and the presence of basal cell carcinoma. Arch Dermatol 137:751–754

67. Kosmadaki MG, Gilchrest BA (2004) The role of telomeres in skin aging/photoaging. Micron 35:155–159

68. Fligiel SE, Varani J, Datta SC, Kang S, Fisher GJ, Voorhees JJ (2003) Collagen degradation in aged/photodamaged skin in vivo and after exposure to matrix metalloproteinase-1 in vitro. J Invest Dermatol 120:842–848

69. Yaar M (1995) Molecular mechanisms of skin aging. Adv Dermatol 10:63–75

70. Khalil Z, Helme R (1996) Sensory peptides as neuromodulators of wound healing in aged rats. J Gerontol A Biol Sci Med Sci 51:B354–361

71. Khalil Z, Ralevic V, Bassirat M, Dusting GJ, Helme RD (1994) Effects of ageing on sensory nerve function in rat skin. Brain Res 641:265–272

72. Gibson SJ, Farrell M (2004) A review of age differences in the neurophysiology of nociception and the perceptual experience of pain. Clin J Pain 20:227–239

73. Wei Q, Matanoski GM, Farmer ER, Hedayati MA, Grossman L (1993) DNA repair and aging in basal cell carcinoma: a molecular epidemiology study. Proc Natl Acad Sci U S A 90:1614–1618

74. Goukassian D, Gad F, Yaar M, Eller MS, Nehal US, Gilchrest BA (2000) Mechanisms and implications of the age-associated decrease in DNA repair capacity. FASEB J 14:1325–1334

75. Elias PM, Ghadially R (2002) The aged epidermal permeability barrier: basis for functional abnormalities. Clin Geriatr Med 18:103–120

76. Jankovic V, Messaoudi I, Nikolich-Zugich J (2003) Phenotypic and functional T-cell aging in rhesus macaques (Macaca mulatta): differential behavior of CD4 and CD8 subsets. Blood 102:3244–3251

77. Dunn-Walters DK, Banerjee M, Mehr R (2003) Effects of age on antibody affinity maturation. Biochem Soc Trans 31:447–448

78. Mouton CP, Bazaldua OV, Pierce B, Espino DV (2001) Common infections in older adults. Am Fam Physician 63:257–268

79. Lemire JM (1995) Immunomodulatory actions of 1,25-dihydroxyvitamin D3. J Steroid Biochem Mol Biol 53:599–602

80. D'Ambrosio D, Cippitelli M, Cocciolo MG, Mazzeo D, Di Lucia P, Lang R, Sinigaglia F, Panina-Bordignon P (1998) Inhibition of IL-12 production by 1,25-dihydroxyvitamin D3. Involvement of NF-kappaB downregulation in transcriptional repression of the p40 gene. J Clin Invest 101:252–262

81. Garland CF, Garland FC, Gorham ED (1999) Calcium and vitamin D. Their potential roles in colon and breast cancer prevention. Ann N Y Acad Sci 889:107–119

82. Lowe L, Hansen CM, Senaratne S, Colston KW (2003) Mechanisms implicated in the growth regulatory effects of vitamin D compounds in breast cancer cells. Recent Results Cancer Res 164:99–110

83. Gloth FM 3rd,Gundberg CM, Hollis BW, Haddad JG Jr, Tobin JD (1995) Vitamin D deficiency in homebound elderly persons. JAMA 274:1683–1686

84. MacLaughlin J, Holick MF (1985) Aging decreases the capacity of human skin to produce vitamin D3. J Clin Invest 76:1536–1538

85. Ghadially R, Brown BE, Sequeira-Martin SM, Feingold KR, Elias PM (1995) The aged epidermal permeability barrier. Structural, functional, and lipid biochemical abnormalities in humans and a senescent murine model. J Clin Invest 95:2281–2290

86. Denda M, Koyama J, Hori J, Horii I, Takahashi M, Hara M, Tagami H (1993) Age- and sex-dependent change in stratum corneum sphingolipids. Arch Dermatol Res 285:415–417

Chapter 3

Ethnic Differences in Skin Aging

3

Erwin Tschachler, Frederique Morizot

Contents

3.1 Introduction

Aging is associated with progressive changes in all tissues including the skin. These changes include a loss of cells and interstitial matrix proteins [1]. In the skin, intrinsic aging involves a reduction of skin thickness and characteristic changes of the tissue architecture [2]. Clinically these changes manifest as wrinkles, tissue slackening, and irregularities of pigmentation, features which, in contrast to aging of other organs, are visible and provide societal clues to estimate the age of the individual. The age at manifestation as well as the severity of the age-associated skin changes differ among individuals and are strongly influenced by environmental factors, particularly life-time UV exposure [3], and also life-style and genetic factors [3, 4]. Phenotypical and functional differences in the skin of individuals of different ethnic backgrounds has attracted considerable interest in dermatology and in the cosmetic industry, and this subject has been excellently reviewed recently by Taylor [5]. The most consistent data from these studies concern the pigmentary system, in particular the size and distribution of melanosomes which govern the different shades of color [5–9]. In contrast to the studies on the pigmentary system, studies on differences in other skin compartments have yielded no definite, and frequently even conflicting, results [5, 8, 10–12].

It is widely assumed that differences in ethnic background influence skin aging. The hallmarks of intrinsic aging such as thinning of the epidermis, flattening of the dermoepidermal junction and reduction of extracellular matrix components [1, 2] apply to all human beings. Similarly, the impact of chronic sun exposure leading to photoaging [3] has also been reported to play an important role, not only in subjects

3

with white skin but also in Asian and black subjects [13–18]. However, the manifestations of photoaging appear to be greatly influenced by natural pigmentation of the skin: they tend to occur at a later age in black subjects than in white subjects [14], and to manifest more with pigment irregularities in both black and Asian subjects than in white subjects [16–18]. Despite the great interest in the subject, comprehensive systematic studies comparing aging and photoaging between different ethnic groups remain to be published. The lack of such studies at present is probably due to the inherent difficulties in designing study protocols that exclude confounding factors in skin aging such as life-style and nutritional habits as well as different environmental influences.

3.2 Comparison of the Occurrence of Age-Associated Changes in the Facial Skin Between French and Japanese Women

A prerequisite for studying the presence and severity of signs of skin aging in populations of different backgrounds is the availability of scales to allow comparison. Until recently, the visible features of photoaging in Asian populations have been contrasted to those in Caucasians and separate grading systems for aging have been developed [18–20], an approach which makes a direct comparison of aging skin between Asians and Caucasians difficult. To avoid as much as possible the distraction of overall facial features that influence the assessment of signs of skin aging, we chose not to use global scales. Instead we established scales documenting different degrees of severity of nine individual visual features of skin aging: expression lines on the forehead, frown lines, pigment spots, crow's feet, wrinkles under the eyes, drooping eyelids, nasolabial folds, wrinkles on the upper lip, and tissue slackening of the lower part of the face. These reference scales were established from a collection of digital images of 138 healthy French Caucasian women aged from 20 to 80 years [21]. Subsequently four dermatologists used these scales to grade aging on digital facial images. The interob-

server agreement ranged from good to very good (kappa values from 0.63 to 0.86), with the notable exception of the features related to skin pigmentation (kappa values from 0.39 to 0.61). Similarly, the intraobserver agreement assessed in two sessions 4 weeks apart ranged from good to very good for the two investigators involved in this part of the study (kappa values ranging from 0.62 to 0.79). When these scales were used to grade skin aging on photographs of Japanese women, the results obtained were very similar to those obtained with the photographs of French women, showing an acceptable degree of interobserver agreement (kappa values >0.60). These results demonstrate that photographic reference scales established in one ethnic population can be used to assess skin aging in individuals of different ethnic backgrounds.

Age-associated skin features have been evaluated primarily by determining either the presence of individual signs or global aging by direct clinical assessment of subjects. This approach has the disadvantage that, apart from the skin changes, other factors such as posture, changes in facial expression and the social interaction between the examiner and the study subject during the examination may influence the examiner's perception and judgment [22]. In addition, the examiner's assessment of a given individual may be influenced by the individual seen before because of the tendency to compare a given state to a previous one.

A possible way to minimize these problems is to analyze age-associated facial skin features from high-quality digital images obtained under standardized lighting and positioning conditions. Using digital images rather than direct clinical examination to investigate aging offers the advantage that the same subject can be studied by several examiners under identical conditions and that the examinations can be easily repeated or even subjected to automated image analysis [17]. One of the main disadvantages concerns the lack of detailed 3D views of the signs of aging. This makes it difficult if not impossible to distinguish between senile lentigines and flat seborrheic keratoses which are common pigmented features in both Asians and Caucasians over 40 years of age [23].

Table 3.1. Characteristics of the study populations (*NS* not significant)

Characteristic	French (*n*=280)	Japanese (*n*=258)	P value (χ² test)
Age (years, mean±SD)	50±17	47±15	NS
Age group (years, number of subjects)			
18–34	65	47	
35–49	64	71	
50–64	79	78	
65–80	72	60	
Self assessment of skin phototype (% of subjects)			
I	8	32	<0.001
II	29	8	
III	43	45	
IV	20	16	
Menopause (% of subjects)	54	39	NS
Regular smoking (% of subjects)	23	16	NS
Self assessment of life-time sun exposure (% of subjects)			
Strong	14	21	NS
Moderate	64	63	
Low or absent	22	15	

In a study to compare the rate of appearance of signs of skin aging in Japanese and French women, images of 256 and 280 subjects, respectively, were taken in Sendai and Paris [21]. The age of the participants ranged from 20 to 80 years. Women were excluded if they had active skin disease, had undergone systemic or local treatments with potential effects on the skin (except oral contraception or hormonal replacement therapy) or had intentionally exposed themselves to the sun during the month prior to the study. The characteristics of the study populations are listed in Table 3.1. The two groups did not differ significantly with regard to age, smoking habits or self-reported lifetime sun exposure. Despite a comparable age range, significantly more French than Japanese women declared themselves to be menopausal (54.2% versus 38.6%, respectively). The high prevalence of self-assessed Fitzpatrick skin type I in the Japanese women (32.5% as compared 8.2% of the French women) was probably due to a misinterpretation of the relationship between the questions regarding the ability to tan and the burning tendency [24, 25].

Grading of the photographs from the two populations using the photographic reference scales revealed several differences in the frequency and severity of the signs of aging. Wrinkles including expression lines on the forehead, frown lines, crow's feet, wrinkles under the eyes, and wrinkles on the upper lip were more pronounced at an earlier age in French than in Japanese women. For example, among women aged 35–49 years, expression lines of grade 3 or more on the forehead, frown lines, lines under the eyes, and crow's feet were seen in 56%, 34%, 23% and 19% of French women and in 17%, 9%, 7% and 8% of Japanese women, respectively. As we have previously reported [26], wrinkles on the upper lip occurred later than the other wrinkles in French women. In the French women they were present at grades 3 and 4 in 38% of women aged 50 to 64 years (Fig. 3.1C). In contrast, they were present at comparable grades in only 10% of the Japanese women (Fig. 3.1C). In fact, the other types of wrinkles also occurred at comparable severity and frequency on average 15 years earlier in French women than in Japanese women in age groups below 65 years. How-

3

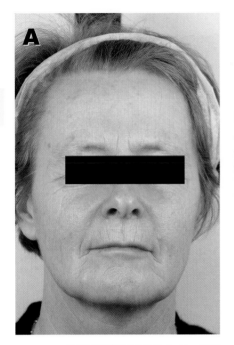

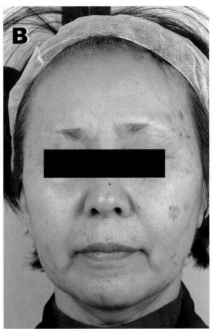

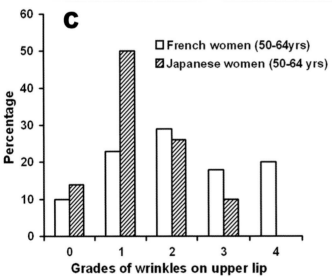

Fig. 3.1. Grading of wrinkles on the upper lip of French and Japanese women. A French woman aged 59 years (**A**) and a Japanese woman aged 64 years (**B**) illustrate the differences in the development of wrinkles in the two populations. Whereas pronounced wrinkles are present on the upper lip of the French woman (grade 4), they are virtually absent in the Japanese woman (grade 2). In a similar way, wrinkles on the other facial areas are less pronounced in the Japanese woman than in the French woman except for the crow's feet which are grade 3 in both women. Note also that the Japanese woman shows grade 3 pigment spots on the forehead and the cheek whereas the French woman shows only grade 2 pigment spots (actinic and seborrheic keratoses are not included in the scales for pigment spots). **C** Grades of wrinkles on the upper lip of French women (*white bars*) and Japanese women (*hatched bars*) women in the age group 50–64 years (written permission for reproduction of the photographs was given by the subjects themselves)

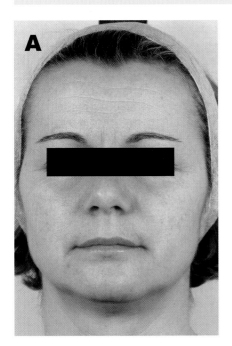

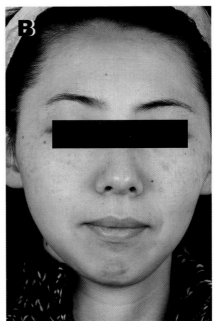

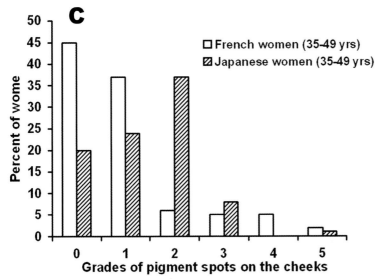

Fig. 3.2. Grading of pigment spots in French and Japanese women. A French woman aged 36 years (**A**) and a Japanese woman aged 39 years (**B**) illustrate the difference in pigment spots in the two populations. Pigment spots are only slightly visible (grade 1) in the French woman whereas they are present at grade 3 on the cheeks of the Japanese woman. Note also that in the Japanese woman wrinkles are either absent or only slightly visible (grade 0 to 1) in all facial areas except the forehead which demonstrates grade 3 wrinkles. In contrast, grade 2 crow's feet and grade 5 expression lines on the forehead are already present in the French woman. **C** Grades of pigment spots on the cheeks of French women (*white bars*) and Japanese women (*hatched bars*) in the age group 35–49 years (written permission for reproduction of the photographs was given by the subjects themselves)

3

ever, above 65 years the differences between the two populations markedly decreased or disappeared altogether.

With regard to signs of tissue slackening, marked nasolabial folds (grade 3 or more) were present in 71% and 95% of French women and 80% and 98% of Japanese women in the age groups 50–64 and 65–80 years, respectively. Loss of the oval of the face and drooping eyelids mostly occurred in women over 50 years of age. There were no substantial differences between French and Japanese women, but lax eyelids were more marked in Japanese than in French women in the age groups 50–64 years (grades 4 and 5 in 28% versus 14%) and 65–80 years (grades 4 and 5 in 90% versus 44.4%).

Early onset of pigment irregularities in Asian women has been reported previously [17–19]. We found that the rates of occurrence and grades of pigment spots differed according to the facial region. Pigment spots of grade 2 or more were observed on the forehead of fewer than 30% of women up to the age of 64 years and in around 50% of women aged between 65 and 80 years with no substantial difference between French and Japanese women. By contrast, pigment spots on the cheeks were more pronounced and occurred earlier than on the forehead, and a distinct difference between French and Japanese women was evident. Whereas at age 35–49 years, only 17% of French women had pigment spots of grade 2 or more, they were present in 46% of Japanese women (Fig. 3.2). This difference between the study samples was maintained through the higher age groups as well (Fig. 1A, B).

3.3 Discussion and Outlook

To expand baseline information on skin aging in different populations, we compared the age of occurrence and the severity of defined signs of skin aging in a cross-sectional study of women in Sendai (Japan) and Paris (France). As far as the technical aspects of studying aging in different ethnic populations are concerned, it was interesting to see that scales for the individual features of aging skin, despite having been developed from images of Caucasian women, showed

acceptable inter- and intraobserver agreement when used to grade skin aging in an Asian population. Our results demonstrate that there are distinct differences in the age of appearance and the rate of deterioration of signs of aging in age-matched groups of women from Japan and France. These differences concern wrinkles as well as age-associated pigment spots. Different types of wrinkles occurred at an earlier age and were more severe in the French women than in the Japanese women. However, two signs of aging associated with tissue slackening, i.e. prominent nasolabial folds and loss of the oval of the face, showed no such marked difference between the two study groups. By contrast, pigment spots occurred at higher grades and earlier in life in Japanese women than in French women, confirming earlier results showing that pigmentation irregularities are a prominent feature in Asian women [17–19].

What conclusions can be drawn from the differences in these signs of skin aging between the two study populations? As discussed above, we are well aware of the fact that a variety of factors contribute to the time of appearance and severity of the signs of skin aging. Certain wrinkles including crow's feet and wrinkles on the upper lip are commonly thought to be associated with photoaging. Therefore the later occurrence of these signs in middle-aged Japanese women might be either due to less sun exposure or to better coping with the effects of the sun. However, the fact that crow's feet are more prevalent in older Japanese women and, as has been reported previously, that pigmentation irregularities are much more prevalent [17–19] argues against this suggestion. We attempted to match the two study populations with regard to age, and no significant differences were apparent between smoking habits and sun exposure behavior. However, a statistically insignificant difference in the percentage of menopausal women emerged. Whether this represents a genuine difference between the two populations or had its origin in the recruiting procedure remains to be determined in future studies.

Furthermore, there are obvious differences in the nutrition habits between Asians and Western populations [30] and these differences may have an impact on the skin aging process.

We have no information on the actual eating habits among our two study populations – a shortcoming that should also be addressed in future interethnic studies. With regard to sun exposure, the questionnaire did not reveal obvious differences. However, after conducting the study we learned that the classification of skin phototypes [24, 25] and the questions regarding reactions to the sun might have been interpreted differently by the Japanese than by the French women. Therefore, we must be cautious in interpreting the impact of sun on skin aging in our Japanese study group. Nevertheless, it has been shown that despite the fact that they are frequently classified as phototype 5, a significant proportion of Asian women experience sunburn [9].

Finally, the two cities where the studies were conducted are not at the same latitude. Sendai (about 38°N) is about 10° closer to the equator than Paris (about 45°N), and this might result in different life-time sun exposures. In this context, Hillebrand et al. [17] compared women from two Japanese cities, Akita (about 39°N) and Kagoshima (31°N). They found that photoaging-associated features occurred several years earlier in women from Kagoshima than in women from Akita. Therefore, the expected impact of the difference in latitude in our study would be that the Japanese women would be affected by sun-related aging effects to a greater degree than the French women. Potentially this could even increase the gap with regard to appearance of the signs of aging between the two populations. Unfortunately, a comparison of long-term UVB levels between the two cities is not available to confirm this possibility. For future studies addressing this subject, it will be necessary to include objective indicators of sun exposure and, most importantly, to match visual data with histological and biochemical data of the skin.

The rapidly progressing economic globalization and ease of travel have greatly contributed to an increased interest in understanding individuals and populations of different cultural and ethnic backgrounds. With regard to skin, dermatologists as well as the cosmetic and pharmaceutical industries share a common interest in defining or excluding differences in skin physiology and pathologies between ethnic groups. For example, it is well established that the incidence of certain skin diseases, most notably that of melanomas and squamous cell carcinomas, differ among Caucasians, Asians and Africans [13]. Whereas these differences can primarily be attributed to the obvious differences in skin pigmentation and to the related UV protection, others, such as differences in the incidence of inflammatory skin diseases, remain to be investigated or even verified [27].

Demographic changes in the industrialized world, which have led to an increased proportion of "older" individuals in the population, have strongly reinforced the interest in age-associated physiological and pathological changes and have raised the question as to whether the aging process differs in individuals of different ethnicity. Few systematic studies addressing this question in relation to the skin have so far been published, and these are based either on small samples or on the evaluation of a single type of wrinkles [28, 29]. Because skin aging is greatly influenced by environmental and life-style factors [3, 4], a comparison of different populations must take into account a great variety of confounding factors, a daunting task which probably deters many investigators.

Thus a truly comprehensive unbiased comparison of even the clinical aspects of skin aging in individuals of different ethnic background is still needed. Apart from the need to extend such studies to ethnic groups other than Caucasians and Asians, it will be necessary to gather more detailed information about the respective environments and life-style habits. It would also be highly desirable to compare migrant populations with resident populations of the same ethnic background. The increasing understanding of genetic factors that may directly and indirectly influence the pace of aging will probably lead to studies at a higher level in which specific genetic markers such as the prevalence of polymorphisms of certain genes important for skin homeostasis are assessed. An example of a step in this direction is a recent study by Rana et al. [31] who have demonstrated that the ARG163GLN polymorphism of the *MC1R* gene, which has been reported to govern the pheomelanin content of the skin, occurs at a high frequency in individuals from eastern and southeastern Asia, whereas it is almost absent in Europeans and altogether lacking in Africans.

3.4 Summary

Ethnic differences in skin properties has attracted great interest in the past, but apart from the visible differences in pigmentation, their assessment has frequently yielded indefinite or conflicting results. Whereas it is self-evident that skin aging affects all human beings regardless of their ethnic background, it is commonly assumed that differences in the rate of appearance of the signs of aging exist between different ethnic groups. As far as premature aging due to sun exposure is concerned, evidence exists that natural photoprotection by melanin and melanosomal dispersion influences both the severity and manifestations of photoaging. Unfortunately, there are only a few studies in which aging and photoaging were compared directly in the context of different racial backgrounds. This is probably, at least in part, due to the fact that for such a study a multitude of confounding factors such as differences in environment, food and life-style habits of the different ethnic groups cannot be easily controlled for. Aware of these constraints, we sought to determine whether defined signs of skin aging appear at comparable severity and rates in Japanese and French women and were able to show that the two study populations differed in the rate of appearance of different types of wrinkles and age-associated pigment changes.

Acknowledgements

We thank Dr. Hachiro Tagami and Sabine Guéhenneux for their help in the conduct of the study, Dr. Christiane Guinot for statistical analysis and Heidemarie Rossiter for critically reading the manuscript.

References

1. Jenkins G (2002) Molecular mechanisms of skin ageing. Mech Ageing Dev 123:801–810
2. Lavker RM (1995) Cutaneous aging: chronologic versus photoaging. In: Gilchrest BA (ed) Photodamage. Blackwell Science, Carlton, pp 123–135
3. Yaar M, Eller MS, Gilchrest BA (2002) Fifty years of skin aging. J Investig Dermatol Symp Proc 7:51–58
4. Kipline D, Davis T, Ostler EL, et al (2004) What can progeroid syndromes tell us about human aging? Science 305:1426–1431
5. Taylor SC (2002) Skin of color: biology, structure, function, and implications for dermatologic disease. J Am Acad Dermatol 46:S41–62
6. Toda K, Pathak MA, Parrish A, Fitzpatrick TB (1972) Alteration of racial differences in melanosome distribution in human epidermis after exposure to ultraviolet light. Nat New Biol 236:143–144
7. Szabo G, Gerald AB, Patnak MA, Fitzpatrick TB (1969) Racial differences in the fate of melanosomes in humane epidermis. Nature 222:1081–1082
8. Montagna W, Carlisle K (1991) The architecture of black and white facial skin. J Am Acad Dermatol 24:929–937
9. Smit NPM, Kolb RM, Lentjes EGWM, et al (1998) Variations in melanin formation by cultured melanocytes from different skin types. Arch Dermatol Res 290:342–349
10. Johnson BL Jr (1998) Differences in skin type. In: Johnson BL Jr, Moy RL, White GM (eds) Ethnic skin: medical and surgical. Mosby, St Louis, pp 3–5
11. Thomson ML (1955) Relative efficiency of pigment and horny layer thickness in protecting the skin of Europeans and Africans against solar ultraviolet radiation. J Physiol 127:236–238
12. Whitmore SE, Sago NJ (2000) Caliper-measured skin thickness is similar in white and black women. J Am Acad Dermatol 42:76–79
13. Halder RM, Ara CJ (2003) Skin cancer and photoaging in ethnic skin. Dermatol Clin 21:725–732
14. Halder RM, Grimes PE, McLaurin CI, et al (1989) Incidence of common dermatoses in a predominantly black dermatology practice. Cutis 32:388
15. Kotrajaras R, Kligman AM (1993) The effect of topical tretinoin on photodamaged facial skin: the Thai experience. Br J Dermatol 129:302–309
16. Halder RM (1998) The role of retinoids in the management of cutaneous conditions in blacks. J Am Acad Dermatol. 39:S98–103
17. Hillebrand GG, Miyamoto K, Schnell B, et al (2001) Quantitative evaluation of skin condition in an epidemiological survey of females living in northern versus southern Japan. J Dermatol Sci 27:S42–S52

18. Chung JH, Lee SH, Youn CS, et al (2001) Cutaneous photodamage in Koreans. Arch Dermatol 137:1043–1051

19. Goh SH (1990) The treatment of visible signs of senescence: the Asian experience. Br J Dermatol 122 [Suppl 35]:105–109

20. Larnier C, Ortonne JP, Venot A, et al (1994) Evaluation of cutaneous photodamage using a photographic scale. Br J Dermatol 130:167–173

21. Morizot F, Guehennneux S, Dheurle S, et al (2004) Do features of skin ageing differ between Asian and Caucasian women? J Invest Dermatol 123:A67

22. Meilgaard M, Civille GV, Carr BT (1999) Sensory evaluation techniques, 3rd edn. CRC Press LLC, Boca Raton, pp 37–41

23. Kwon OS, Hwang EJ, Bae JH, et al (2003) Seborrheic keratosis in the Korean males: causative role of sunlight. Photodermatol Photoimmunol Photomed 19:73–80

24. Park SB, Suh DH, Youn JI (1998) Reliability of self-assessment in determining skin phototype of Korean brown skin. Photodermatol Photoimmunol Photomed 14:160–163

25. Stanford DG, Georgouras. KE, Sullivan EA, Greenoak GE (1996) Skin phototyping in Asian Australians. Australasian J Dermatol 37:36–38

26. Guinot C, Malvy DJ, Ambroisine L, et al (2002) Relative contribution of intrinsic vs extrinsic factors to skin aging as determined by a validated skin age score. Arch Dermatol 138:1454–1460

27. Williams H, Robertson C, Stewart A, et al (1999) Worldwide variations in the prevalence of symptoms of atopic eczema in the International Study of Asthma and Allergies in Childhood. J Allergy Clin Immunol 103:125–138

28. Tsukahara K, Fujimura T, Yoshida Y, et al (2004) Comparison of age-related changes in wrinkling and sagging of the skin in Caucasian females and in Japanese females. J Cosmet Sci 55:373–385

29. Hillebrand GG, Levine MJ, Miyamoto K (2001) The age-dependent changes in skin condition in African Americans, Asian Indians, Caucasians, East Asians, and Latinos. IFSCC Magazine 4(4):259–266

30. Park SY, Paik HY, Skinner JD, Spindler AA, Park HR (2004) Nutrients intake of Korean-American, Korean and American adolescents. J Am Diet Assoc 104:242–245

31. Rana BK, Hewett-Emmett D, Jin L, et al (1999) High polymorphism at the human melanocortin 1 receptor locus. Genetics 151:1547–1557

Chapter 4

Photoaging of Skin

Jean Krutmann, Barbara A. Gilchrest

4

Contents

4.1 Introduction

For decades it has been appreciated that aging is a consequence of both genetic and environmental influences. Genetic factors are evident, for example, in the >100-fold variation among species in the rate of aging; and recent studies of fruit flies, worms and even mice have identified specific longevity genes whose modification can greatly alter life-span [27]. Conversely, a role for environmental factors can be deduced both from epidemiological and laboratory-based experimental data. Such influences include ionizing radiation, severe physical and psychological stress, over-eating versus caloric restriction, and in the case of skin ultraviolet (UV) radiation.

Among all environmental factors, solar UV radiation is the most important in premature skin aging, a process accordingly also termed photoaging. Over recent years substantial progress has been made in elucidating the underlying molecular mechanisms. From these studies it is now clear that both UVB (290–320 nm) and UVA (320–400 nm) radiation contribute to photoaging. UV-induced alterations at the level of the dermis are best studied and appear to be largely responsible for the phenotype of photoaged skin. It is also generally agreed that UVB acts preferentially on the epidermis where it not only damages DNA in keratinocytes and melanocytes but also causes the production of soluble factors including proteolytic enzymes which then in a second step affect the dermis; in contrast UVA radiation penetrates far more deeply on average and hence exerts direct effects on both the epidermal and the dermal compartment (Fig. 4.1). UVA is also 10–100 times more abundant in sunlight than UVB, depending on the season and time of day. It has therefore been

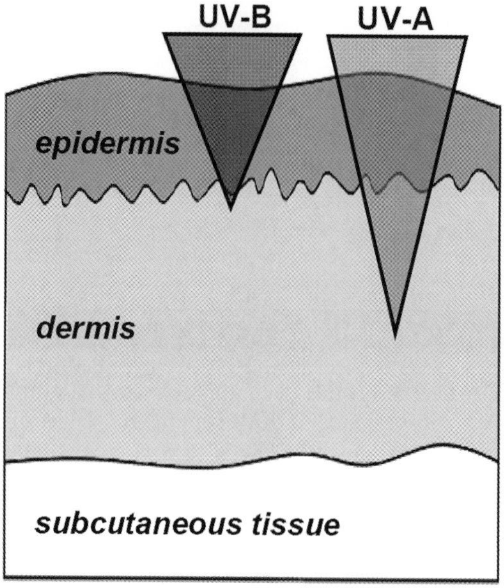

4

Fig. 4.1. Wavelength-dependent penetration of UV radiation into human skin

proposed that, although UVA photons are individually far less biologically active than UVB photons, UVA radiation may be at least as important as UVB radiation in the pathogenesis of photoaging [3].

The exact mechanisms by which UV radiation causes premature skin aging is not yet clear, but a number of molecular pathways explaining one or more of the key features of photoaged skin have been described. Some of these models are based on irradiation protocols which use single or few UV exposures, whereas others take into account the fact that photoaging results from chronic UV damage, and as a consequence employ chronic repetitive irradiation protocols. Still others rely on largely theoretical constructs rather than experimental observations.

4.2 Mechanisms of Photoaging

All organ systems are affected by aging processes, many in organ-specific ways, but aging uniformly has the effect of reducing maximal function and reserve capacity, as well as the ability to compensate for injury and a hostile environ-

ment. Ultimately, such losses are incompatible with life.

Of interest, most if not all age-accelerating environmental factors damage DNA either directly or indirectly, often through oxidation [35]. Furthermore, the rate of aging in various species correlates inversely with the rate and fidelity of DNA repair [25], and most progeroid syndromes for which the genetic lesion have been identified have impaired DNA replication and/or DNA damage responses [35]. In combination with the fact that cumulative DNA damage accompanies chronological aging [44], these observations suggest that both the indisputable heritable component and the environmental component of aging result in large part from changing DNA status during the individual's life. Recent insights into the structure and maintenance of nuclear and mitochondrial DNA (mtDNA) have allowed the development of a unifying hypothesis for the apparently independent processes of "intrinsic" (genetic) and "extrinsic" (environmental) aspects of aging.

The next two sections of this chapter develop this intellectual framework and relate it to the phenomenon of skin aging, and particularly photoaging, by focusing on telomeres, the terminal portion of eukaryotic chromosomes, and on mtDNA [3, 4]). The subsequent sections provide detailed information now available with regard to specific aging targets and signaling pathways responsible for photoaging-associated morphological and functional changes in skin. These include UV-induced alterations of connective tissue components, vascularization patterns, inflammatory cells and protein oxidation. Finally, we present a unifying concept that reconciles the most recent findings in an attempt to provide a novel and comprehensive model to explain photoaging and a framework for future investigations.

4.3 Role of Telomeres in Skin Aging/Photoaging

Telomeres are tandem repeats of a short sequence, in all mammals TTAGGG and its complement, extending for several thousand base pairs at the end of each chromosome [23]. Telomeres do not encode genes, but rather protect the proximal genes and regulatory sequences in several ways. First, in the absence of telomeres, chromosome ends are recognized in cells as double-strand breaks, which lead to inappropriate end-to-end fusions, incompatible with normal cell function [8]. Telomeres prevent the occurrence of such events. Second, at least in part because of the "end replication problem" [33], the final 100–200 bases of the chromosome are not replicated when cells divide. Telomeres thus provide a buffer zone, preventing loss of genetic information in the daughter cells. Third, when telomeres reach a critically short length, after approximately 60 postnatal doublings, cells will no longer divide, regardless of environmental signals – a state termed proliferative senescence [24]. By serving in this way as the "biological clock," telomeres are thought to exert an anticancer effect, preventing indefinite serial division of cells that on a purely statistical basis may have accumulated dangerous mutations over the individual's life-span [12]. Finally, at least under experimental conditions and presumably also in response to physiological challenges, disruption of the normal telomere structure leads to multiple DNA damage responses capable of removing severely compromised cells from the tissue through apoptosis or senescence [8].

Telomeres form a loop structure at chromosome ends, held in place by an overhang of approximately 100–400 bases on the 3′ strand, tandem repeats of TTAGGG that inserts into the proximal double-stranded telomere and is secured there by binding proteins [42]. Considerable data (reviewed in Ref. 21) suggest that exposure of the 3′ overhang occurs at the time of acute DNA damage or when the telomere has become critically short and the loop correspondingly "tight", and that the exposed TTAGGG repeat sequence is then recognized by a sensor protein that initiates DNA damage signaling. This scenario, although still not experimentally confirmed in all its details, offers an attractive explanation for the striking similarities between chronological aging, with its large component of cellular senescence, and the "accelerated aging" that results from environmental insults such as DNA-damaging UV irradiation (Fig. 4.2). Of note, this model is not mutually exclusive with

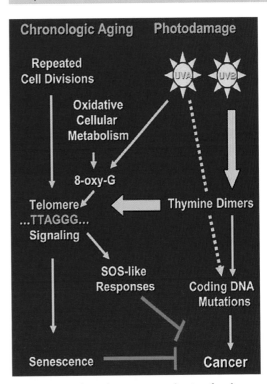

Fig. 4.2. Hypothetical common mechanism for chronological aging and photoaging. Photodamage leads to thymine dimers and (6-4) photoproducts as well as ROS that damage genomic DNA and may give rise to mutations in coding or regulatory DNA sequences of critical genes, eventually leading to the development of cancer. UV also damages telomeres, indeed disproportionately, due to their greater proportion of thymidine dinucleotides (TT) and guanine (G) bases compared to the chromosome overall. Such damage or its attempted repair is postulated to disrupt the telomere loop, expose the TTAGGG overhang, and promote "aging" as described in the text. Chronological (intrinsic) aging is accompanied in most cases by repeated cell divisions that shorten telomeres and may ultimately also destabilize the telomere loop. This process may be accelerated by the compensatory rounds of cell division that follow photodamage. Oxidative cellular metabolism over the life-span also damages DNA generally and telomeres preferentially at guanine residues, providing a further overlap between the mechanisms of aging and acute photodamage, principally UVA photodamage. Exposure of the TTAGGG overhang sequence appears to initiate signaling that, depending on the intensity and duration of the stimulus, may lead to SOS-like responses such as increased melanogenesis, or to proliferative senescence or apoptosis, all of which work against carcinogenesis

4

the other proposed contributors to photoaging discussed below.

One observation suggesting a common molecular mechanism for chronological aging and photoaging is that the same nuclear proteins that mediate senescence after many rounds of cell division also appear to mediate DNA damage signaling after UV irradiation or oxidative stress [46]. Furthermore, multiple exposures of cultured cells to either UV or to an agent that exclusively generates reactive oxygen species (ROS) results in premature entry into the senescent state [45]. Moreover, mutation of any of the several genes that encode proteins implicated in telomere-initiated DNA damage signaling cause diseases characterized by accelerated aging and/ or a high cancer risk. These include ATM (ataxia telangiectasia-mutated), ATR (ATM-related, Sickle syndrome), the protein WRN (mutated in Werner syndrome), the p53 tumor suppressor protein (Li-Fraumeni syndrome), and many others [35]. Indirect experimental evidence that exposure of the telomere 3′ overhang (repeats of TTAGGG approximately 100–400 bases in length [39]) plays a role in both chronological aging and acute or chronic photodamage derives from the demonstration that telomere homolog oligonucleotides, termed T-oligos, when provided to cultured skin-derived cells or topically applied to intact skin produce the same changes. For example, thymidine dinucleotides representing one-third of the TTAGGG repeat sequence, as well as an 11-base 100% homolog, induce tanning in rodent skin and in cultured human melanocytes and skin explants that is indistinguishable from UV-induced tanning following a single UV exposure [16] or from the essentially irreversible hyperpigmentation sometimes termed solar bronzing observed in repeatedly UV-exposed photodamaged skin.

In human melanocytes, disruption of the telomere loop with exposure of the 3′ overhang also induces melanogenesis or tanning [17]. Treatment of dermal fibroblasts with T-oligos similarly mimics acute and chronic UV exposure by inducing metalloproteins, a phenotype also observed in serially passaged non-sun-exposed senescent fibroblasts (unpublished observation). Likewise, β-galactosidase positivity, a poorly understood but reliable indicator of

cellular senescence as well as of chronologically aged and photoaged skin [15], is also observed in human dermal fibroblasts treated with T-oligos or caused to undergo telomere disruption by exposure of the 3′ overhang [34]. Finally, although the ability of UV irradiation to accelerate telomere shortening has not yet been examined, oxidative stress (known to mediate UVA damage in large part) is known to have this effect [45]; and T-oligos also act through ROS to accomplish their cellular effects, including induction over time of senescence (β-galactosidase positivity) in human dermal fibroblasts [31].

4.4 Mitochondrial DNA Mutations and Photoaging

Mitochondria are organelles whose main function is to generate energy for the cell. This is achieved by a multistep process called oxidative phosphorylation or electron transport chain. Located at the inner mitochondrial membrane are five multiprotein complexes that generate an electrochemical proton gradient used in the last step of the process to turn ADP and organophosphate into ATP. This process is not completely error-free and ultimately leads to the generation of ROS, making the mitochondrion the site of the highest ROS turnover in the cell. In close proximity to this site lies the mitochondrion's own genomic material, the mtDNA. Human mtDNA is a 16,559-bp long circular double-stranded molecule of which four to ten copies exist per cell. Since mitochondria do not exhibit any repair mechanism to remove bulky DNA lesions, although they do exhibit a base excision repair mechanism and repair mechanisms against oxidative damage, the mutation frequency of mtDNA is approximately 50-fold higher than that of nuclear DNA.

Mutations of mtDNA have been found to play a causative role in degenerative diseases such as Alzheimer's disease, chronic progressive external ophthalmoplegia and Kearns-Sayre syndrome [14]. In addition to degenerative diseases, mutations of mtDNA may play a causative role in the normal aging process with an accumulation of mtDNA mutations accompanied by a decline of mitochondrial functions [47]. Recent evidence indicates that mtDNA mutations are also involved in the process of photoaging [3].

Photoaged skin is characterized by increased mutations of the mitochondrial genome [1, 7, 49]. Intraindividual comparison studies have revealed that the so-called common deletion, a 4,977-bp deletion of mtDNA, is increased up to tenfold in photoaged skin as compared with sun-protected skin of the same individuals. The amount of the common deletion in human skin does not correlate with chronological aging [29], and it has therefore been proposed that mtDNA mutations such as the common deletion represent molecular markers for photoaging. In support of this concept it has been shown that repetitive, sublethal exposure to UVA radiation at doses that may be acquired during a regular summer holiday induces mutations of mtDNA in cultured primary human dermal fibroblasts in a singlet oxygen-dependent fashion [2]. Even more importantly, in vivo studies have revealed that repetitive exposure three times daily of previously unirradiated buttock skin for a total of 2 weeks to physiological doses of UVA radiation leads to an approximately 40% increase in the levels of the common deletion in the dermal, but not the epidermal, compartment of irradiated skin [4]. Furthermore, it has been shown that, once induced, these mutations persist for at least 16 months in UV-exposed skin. Interestingly, in a number of individuals, the levels of the common deletion in irradiated skin continued to increase with a magnitude up to 32-fold.

It has been postulated for the normal aging process as well as for photoaging that the induction of ROS generates mtDNA mutations, in turn leading to a defective respiratory chain and, in a vicious cycle, inducing even more ROS and subsequently allowing mtDNA mutagenesis independent of the inducing agent [26]. It is the characteristic of vicious cycles that they evolve at ever-increasing speeds. Thus, the increase of the common deletion up to levels of 32-fold, independent of UV exposure, may represent the first in vivo evidence for the presence of such a vicious cycle in general and in human skin in particular (Fig. 4.3).

The mechanisms by which generation of mtDNA mutations by UVA exposure translates into the morphological alterations observed in photoaging human skin are currently being unraveled. In general, a cause–effect relationship between premature aging and mtDNA mutagenesis is strongly suggested by studies employing homozygous knock-in mice that express a proofreading-deficient version of PolyA, the nucleus-encoded subunit of mtDNA polymerase [41]. As expected, these mice develop a mtDNA mutator phenotype with increased amounts of deleted mtDNA. This increase in somatic mtDNA mutations has been found to be associated with a reduced life-span and premature onset of aging-related phenotypes such as weight loss, reduced subcutaneous fat, alopecia, kyphosis, osteoporosis, anemia, reduced fertility and heart enlargement.

In addition, recent studies have demonstrated that UVA radiation-induced mtDNA mutagenesis is of functional relevance in primary human dermal fibroblasts and apparently has molecular consequences suggestive of a causative role for mtDNA mutations in photoaging of human skin as well [5]. Accordingly, induction of the common deletion in human skin fibroblasts is paralleled by a measurable decrease of oxygen consumption, mitochondrial membrane potential and ATP content as well as an increase of MMP-1, while tissue-specific inhibitors of MMPs (TIMPs) remain unaltered, an imbalance that is known to be involved in photoaging of human skin (see below). These observations suggest a link not only between mutations of mtDNA and cellular energy metabolism, but also between mtDNA mutagenesis, energy metabolism and a fibroblast gene expression profile that would functionally correlate with increased matrix degradation and thus premature skin aging. In order to provide further evidence for a role of the energy metabolism in mtDNA mutagenesis and the development of this "photoaging phenotype", the effect of creatine was studied in these cells. This tested the hypothesis that generation of phosphocreatine, and consequently ATP, is facilitated if creatine is abundant in cells. This would allow easier binding of existing energy-rich phosphates to the energy precursor creatine. Indeed, experimental supplementation of normal human fibroblasts with creatine normalizes mitochondrial mutagenesis as well as the functional parameters oxygen

4

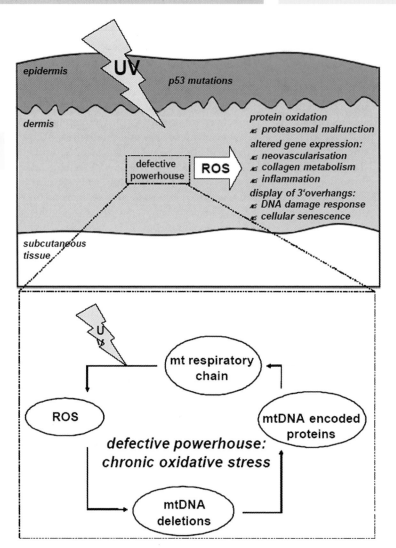

Fig. 4.3. The "defective powerhouse model" of cutaneous aging. In human dermal fibroblasts, the persistence of UV-radiation-induced mtDNA mutations and the resulting vicious cycle (*inset*) leads to inadequate energy production and chronic oxidative stress (i.e. the defective powerhouse). As a consequence of this situation, (1) the repetitive display of the 3′ overhang of telomeres initiates a DNA damage response and ultimately cellular senescence, (2) the gene expression pattern of dermal fibroblasts is altered in a way that induces neovascularization, disturbs collagen metabolism and causes the generation of an inflammatory infiltrate, and (3) intracellular proteins are being oxidized, which eventually leads to inhibition of the proteasome with functional consequences that manifest extracellularly. In addition, UV radiation directly (without the induction of mtDNA mutations) causes functional and structural alterations within the dermis and epidermis (e.g. p53 mutations) which contribute to premature skin aging

consumption and MMP-1 while an inhibitor of creatine uptake abrogates this effect [5].

4.5 Connective-tissue Alterations in Photoaging: the Role of Matrix Metalloproteinases and Collagen Synthesis

Photoaged skin is characterized by alterations to the dermal connective tissue. The extracellular matrix in the dermis mainly consists of type I and type III collagen, elastin, proteoglycans, and fibronectin. In particular, collagen fibrils are important for the strength and resilience of skin, and alterations in their number and structure are thought to be responsible for wrinkle formation.

In photoaged skin, collagen fibrils are disorganized and abnormal elastin-containing material accumulates [38]. Biochemical studies have revealed that in photoaged skin levels of types I and III collagen precursors and crosslinks are reduced, whereas elastin levels are increased [10, 40].

How does UV radiation cause these alterations? In principle it is reasonable to assume that UV radiation leads to an enhanced and accelerated degradation and/or a decreased synthesis of collagen fibers, and our current knowledge indicates that both mechanisms may be involved.

A large number of studies unambiguously demonstrate that the induction of matrix metalloproteinases (MMPs) plays a major role in the pathogenesis of photoaging. As indicated by their name, these zinc-dependent endopeptidases show proteolytic activity in their ability to degrade matrix proteins such as collagen and elastin. Each MMP degrades different dermal matrix proteins; for example, MMP-1 cleaves collagen types I, II and III, whereas MMP-9, which is also called gelatinase, degrades collagen types IV and V and gelatin. Under basal conditions, MMPs are part of a coordinated network and are precisely regulated by their endogenous inhibitors, i.e. TIMPs, which specifically inactivate certain MMPs. An imbalance between activation of MMPs and their respective TIMPs could lead to excessive proteolysis.

It is now very well established that UV radiation induces MMPs without affecting the expression or activity of TIMPs [19, 37]. These MMPs can be induced by both UVB and UVA radiation, but the underlying photobiological and molecular mechanisms differ depending on the type of irradiation. In a very simplified scheme, UVA radiation would mostly act indirectly through the generation of ROS, in particular singlet oxygen, which subsequently can exert a multitude of effects such as lipid peroxidation, activation of transcription factors and generation of DNA strand breaks [37]. While UVB radiation-induced MMP induction has been shown to involve the generation of ROS as well [48], the main mechanism of action of UVB is by direct interaction with DNA via the induction of DNA damage. Recent studies have indeed provided evidence that enhanced repair of UVB-induced cyclobutane pyrimidine dimers in the DNA of epidermal keratinocytes through topical application of liposomally encapsulated DNA repair enzymes on UVB-irradiated human skin prevents UVB radiation-induced epidermal MMP expression (Yarosh D, Krutmann J, unpublished observation).

The activity of MMPs is tightly regulated by transcriptional regulation and elegant in vivo studies by Fisher et al. have demonstrated that exposure of human skin to UVB radiation leads to the activation of the respective transcription factors [18]. Accordingly, UV exposure of human skin not only leads to the induction of MMPs within hours after irradiation, but already within minutes, transcription factors AP-1 and NFkB, which are known stimulatory factors of MMP genes, are induced. These effects can be observed at low UVB dose levels, because transcription factor activation and MMP-1 induction can be achieved by exposing human skin to one-tenth of the dose necessary for skin reddening (0.1 minimal erythema dose). Subsequent work by the same group clarified the major components of the molecular pathway by which UVB exposure leads to the degradation of matrix proteins in human skin. Low-dose UVB irradiation induces a signaling cascade which involves upregulation of epidermal growth factor receptors (EGFR), the GTP-binding regulatory protein p21Ras, extracellular signal-regu-

4

lated kinase (ERK), c-jun aminoterminal kinase (JNK) and p38. Elevated c-jun together with constitutively expressed c-fos increases activation of AP-1. Identification of this UVB-induced signaling pathway not only unravels the complexity of the molecular basis which underlies UVB radiation-induced gene expression in human skin, but also provides a rationale for the efficacy of tretinoin (all-*trans*-retinoic acid) in the treatment of photoaged skin. Accordingly, topical pretreatment with tretinoin inhibits the induction and activity of MMPs in UVB-irradiated skin through prevention of AP-1 activation.

In addition to destruction of existing collagen through activation of MMPs, failure to replace damaged collagen is thought to contribute to photoaging as well. Accordingly, in chronically photodamaged skin, collagen synthesis is downregulated as compared to sun-protected skin [20]. The mechanism by which UV radiation interferes with collagen synthesis is not yet known but a recent study has provided evidence that fibroblasts in severely (photo)damaged skin have less interaction with intact collagen and are thus exposed to less mechanical tension, and it has been proposed that this situation might lead to decreased collagen synthesis [43].

4.6 UV-Induced Modulation of Vascularization

There is increasing evidence that cutaneous blood vessels may play a role in the pathogenesis of photoaging. Photoaged skin shows vascular damage intrinsically aged skin does not. In mildly photodamaged skin, there is venular wall thickening, while in severely damaged skin the vessel walls are thinned and the supporting perivascular veil cells are reduced in number [11]. The number of vascular cross-sections is reduced [28] and there are local dilations, corresponding to clinical telangiectases. Overall, there is a marked change in the horizontal vascularization pattern with dilated and distorted vessels. Studies in humans and in the hairless skh-1 mouse model for skin aging have demonstrated that acute and chronic UVB irradiation greatly increases skin vascularization [6, 50].

The formation of blood vessels from preexisting vessels is tightly controlled by a number of angiogenic factors and factors which inhibit angiogenesis. These growth factors include basic fibroblast growth factor, interleukin-8, tumor growth factor-beta, platelet-derived growth factor and vascular endothelial growth factor (VEGF). VEGF appears to be involved in chronic UVB damage because UVB radiation-induced dermal angiogenesis in Skh-1 mice is associated with increased VEGF expression in the hyperplastic epidermis of these animals [50]. Even more importantly, targeted overexpression of the angiogenesis inhibitor thrombospondin-1 not only prevents UVB radiation-induced skin vascularization and endothelial cell proliferation, but also significantly reduces dermal photodamage and wrinkle formation. These studies suggest that UVB radiation-induced angiogenesis plays a direct biological role in photoaging.

4.7 Photoaging as a Chronic Inflammatory Process

In contrast to intrinsically aged skin, which shows an overall reduction in cell numbers, photoaged skin is characterized by an increase in the number of dermal fibroblasts, which appear hyperplastic, but also by increased numbers of mast cells, histiocytes and mononuclear cells. The presence of such a dermal infiltrate indicates the possibility that a chronic inflammatory process takes place in photoaged skin, and in order to describe this situation the terms heliodermatitis and dermatoheliosis have been coined [30]. More recent studies have shown that increased numbers of CD4[+] T-cells are present in the dermis whereas intraepidermally, infiltrates of indeterminate cells and a concomitant reduction in the number of epidermal Langerhans cells have been described [13, 22]. It is currently not known whether the presence of inflammatory cells represents an epiphenomenon or whether these cells play a causative role in the pathogenesis of photoaging, e.g. through the production of soluble mediators which could affect the production and/or degradation of extracellular matrix proteins.

4.8 Protein Oxidation and Photoaging

The aging process is accompanied by enhanced oxidative damage. All cellular components including proteins are affected by oxidation [32]. Protein carbonyls may be formed either by oxidative cleavage of proteins or by direct oxidation of lysine, arginine, proline, and threonine residues. In addition, carbonyl groups may be introduced into proteins by reaction with aldehydes produced during lipid peroxidation or with reactive carbonyl derivatives generated as a consequence of the reaction with reducing sugars or their oxidation products with lysine residues of proteins.

Within the cell, the proteasome is responsible for the degradation of oxidized proteins. During the aging process this function of the proteasome is diminished and oxidized proteins accumulate. In addition, lipofuscin, a highly crosslinked and modified protein aggregate, is formed. This aggregate accumulates within cells and is able to inhibit the proteasome. These alterations mainly occur within the cytoplasm and lipofuscin does not accumulate in the nucleus.

In biopsies from individuals with histologically confirmed solar elastosis, an accumulation of oxidatively modified proteins was found specifically within the upper dermis [36]. Protein oxidation in photoaged skin is most likely due to UV irradiation because repetitive exposure of human buttock skin over 10 days to increasing UV doses as well as in vitro irradiation of cultured dermal fibroblasts with UVB or UVA has been shown to cause protein oxidation. The functional relevance of increased protein oxidation in UV-irradiated dermal fibroblasts, in particular with regard to the pathogenesis of photoaging, is currently not known. Very recent studies, however, indicate that increased protein oxidation, that may result from a single exposure of cultured human fibroblasts to UVA radiation, inhibits proteasomal functions and thereby affects intracellular signaling pathways which are involved in MMP-1 expression (Krutmann J et al., unpublished data).

4.9 Concluding Remarks: Towards a Unifying Concept

From the above it is evident that major progress has been made recently in identifying molecular mechanisms involved in photoaging. In this regard skin has proven to serve as an excellent model organ to understand basic mechanisms relevant to extrinsic aging.

Despite all this progress, however, a general, unifying concept linking the different mechanisms and molecular targets described in the previous paragraphs is still lacking. In other words, the critical question to answer is: How do mtDNA mutagenesis, telomere shortening, neovascularization, protein oxidation, downregulation of collagen synthesis, and increased expression of matrix metalloproteinases together cause photoaging of human skin? Which of these mechanisms are of primary importance and responsible for inducing others? Are some or all of the above-mentioned characteristics of photoaged skin merely epiphenomena and, if so, to what extent are they causally related to premature skin aging?

The current state of knowledge does not allow us to answer these questions in a definitive manner. Nevertheless, we would like to propose a hypothesis which tries to reconcile most of the research discussed above in one model.

By definition, photoaging is the superposition of chronic UV damage on the intrinsic aging process. We envision photoaging of human skin to include both accelerated telomere shortening and UV radiation-induced mtDNA mutagenesis in the dermis of human skin. We believe that the persistence of UV radiation-induced mtDNA mutations and the resulting vicious cycle with further increases in mtDNA mutations leads to a situation which can best be described as a "defective powerhouse" where inadequate energy production leads to chronic oxidative stress (Fig. 4.3). In the epidermis, direct UVB-induced DNA damage in combination with indirect ROS-induced damage would be expected to cause the well-documented "UV signature mutations" in p53 [9], leading to poorly regulated growth and differentiation of epidermal cells associated both with discrete premalignant

4

actinic keratosis and diffuse photoaging. In the dermis, functional consequences of direct DNA damage and aberrant ROS production in human dermal fibroblasts could be: (1) the repetitive display of the 3′ overhang of the telomeres, the subsequent initiation of a DNA damage response and ultimately the induction of cellular senescence, (2) an altered gene expression pattern which would affect neovascularization and collagen metabolism and possibly also the generation of an inflammatory infiltrate, and (3) the oxidation of intracellular proteins and inhibition of the proteasome.

Evidence supporting this model is now being generated in cultured skin-derived cells. However, development of appropriate mouse models characterized by different susceptibilities towards UV-radiation-induced mtDNA mutagenesis as well as telomere shortening and subsequent DNA damage signaling is of the utmost importance.

References

1. Berneburg M, Gattermann N, Stege H, Grewe M, Vogelsang K, Ruzicka T, Krutmann J (1997) Chronically ultraviolet-exposed human skin shows a higher mutation frequency of mitochondrial DNA as compared to unexposed skin and the hematopoietic system. Photochem Photobiol 66:271–275
2. Berneburg M, Grether-Beck S, Kurten V, Ruzicka T, Briviba K, Sies H, Krutmann J (1999) Singlet oxygen mediates the UVA-induced generation of the photoaging-associated mitochondrial common deletion. J Biol Chem 274:15345–15349
3. Berneburg M, Plettenberg H, Krutmann J (2000) Photoaging of human skin. Photodermatol Photoimmunol Photomed 16:239–244
4. Berneburg M, Plettenberg H, Medve-Konig K, Pfahlberg A, Gers-Barlag H, Gefeller O, Krutmann J (2004) Induction of the photoaging-associated mitochondrial common deletion in vivo in normal human skin. J Invest Dermatol 122:1277–1283
5. Berneburg M, Gremmel T, Kurten V, Schroeder P, Hertel I, von Mikecz A, Wild S, Chen M, Declercq L, Matsui M, Ruzicka T, Krutmann J (2005) Creatine supplementation normalizes mutagenesis of mitochondrial DNA as well as functional consequences. J Invest Dermatol 125:213–220
6. Bielenberg DR, Bucana CD, Sanchez R, Donawho CK, Kripke ML, Fidler IJ (1998) Molecular regulation of UVB-induced angiogenesis. J Invest Dermatol 111:864–872
7. Birch-Machin MA, Tindall M, Turner R, Haldane F, Rees JL (1998) Mitochondrial DNA deletions in human skin reflect photo- rather than chronologic aging. J Invest Dermatol 110:149–152
8. Blackburn EH (2001) Switching and signaling at the telomere. Cell 106:661–673
9. Brash D, Rudolph J, Simon J, Lin A, McKenna G, Baden H, Halperin A, Pontenm J (1991) A role for sunlight in skin cancer: UV-induced p53 mutations in squamous cell carcinoma. Proc Natl Acad Sci U S A 88:10124–10128
10. Braverman IM, Fonferko E (1982) Studies in cutaneous aging: I. The elastic fibre network. J Invest Dermatol 78:434–443
11. Braverman IM, Fonferko E (1982) Studies in cutaneous aging: II. The microvasculature. J Invest Dermatol 78:444–448
12. Campisi J (2001) Cellular senescence as a tumor-suppressor mechanism. Trends Cell Biol 11:S27–31
13. DeLeo VA, Dawes L, Jackson R (1981) Density of Langerhans cells (LC) in normal versus chronic actinically damaged skin (CADS) of humans. J Invest Dermatol 76:330–334
14. DiMauro S, Schon EA (2003) Mitochondrial respiratory-chain diseases. N Engl J Med 348:2656–2668
15. Dimri GP, Lee X, Basile G, Acosta M, Scott G, Roskelley C, Medrano EE, Linskens M, Rubelj I, Pereira-Smith O, et al (1995) A biomarker that identifies senescent human cells in culture and in aging skin in vivo. Proc Natl Acad Sci U S A 92:9363–9367
16. Eller MS, Yaar M, Gilchrest BA (1994) DNA damage and melanogenesis. Nature 372:413–414
17. Eller MS, Liou J, Gilchrest BA (2003) Disruption of the telomere loop induces pigmentation in normal human melanocytes. J Invest Dermatol 121
18. Fisher GJ, Wang ZQ, Datta SC, Varani J, Kang S, Voorhees JJ (1997) Pathophysiology of premature skin aging induced by ultraviolet light. N Engl J Med 337:1419–1428
19. Fisher GJ, Talwar HS, Lin J, et al (1998) retinoic acid inhibits induction of c-jun protein by ultraviolet radiation that occurs subsequent to activation of mitogen-activated protein kinase pathways in human skin in vivo. J Clin Invest 101:1432–1440
20. Fisher G, Datta S, Wang Z, Li X, Quan T, Chung J, Kang S, Voorhees J (2000) c-Jun dependent inhibition of cutaneous procollagen transcription following ultraviolet irradiation is reversed by all-trans retinoid acid. J Clin Invest 106:661–668
21. Gilchrest BA (2004) Using DNA damage responses to prevent and treat skin cancers. J Dermatol 31:862–877

22. Gilchrest BA, Murphy GF, Soter NA (1982) Effects of chronologic aging and ultraviolet irradiation on Langerhans cells in human skin. J Invest Dermatol 79:85–88

23. Greider CW (1996) Telomere length regulation. Annu Rev Biochem 65:337–65

24. Harley CB, Futcher AB, Greider CW (1990) Telomeres shorten during ageing of human fibroblasts. Nature 345:458–60

25. Hart RW, Setlow RB (1974) Correlation between deoxyribonucleic acid excision-repair and life-span in a number of mammalian species. Proc Natl Acad Sci U S A 71:2169–73

26. Jacobs HT (2003) The mitochondrial theory of aging: dead or alive? Aging Cell 2:11–7

27. Kenyon C (2005) The plasticity of aging: insights from long-lived mutants. Cell 120:449–60

28. Kligman AM (1979) Perspectives and problems in cutaneous gerontology. J Invest Dermatol 73:39–46

29. Koch H, Wittern KP, Bergemann J (2001) In human keratinocytes the Common Deletion reflects donor variabilities rather than chronologic aging and can be induced by ultraviolet A irradiation. J Invest Dermatol 117:892–7

30. Lavker RM, Kligman A (1988) Chronic heliodermatitis: a morphologic evaluation of chronic actinic dermal damage with emphasis on the role of mast cells. J Invest Dermatol 90:325–330

31. Lee-Bellantoni MS (2005) Antioxidant defense and redox responses to telomere homolog oligonucleotides in human dermal fibroblasts: A model for investigating redox signaling responses to DNA damage. Department of Pathology & Laboratory Medicine, Division of Graduate Medical Science. Boston: Boston University School of Medicine

32. Levine RL, Stadtman ER (2001) Oxidative modification of protein during ageing. Exp Gerontol 36:1495–1502

33. Levy MZ, Allsopp RC, Futcher AB, Greider CW, Harley CB (1992) Telomere end-replication problem and cell aging. J Mol Biol 225:951–60

34. Li GZ, Eller MS, Firoozabadi R, Gilchrest BA (2003) Evidence that exposure of the telomere 3′ overhang sequence induces senescence. Proc Natl Acad Sci U S A 100:527–31

35. Lombard DB, Chua KF, Mostoslavsky R, Franco S, Gostissa M, Alt FW (2005) DNA repair, genome stability, and aging. Cell 120:497–512

36. Sander CS, Chang H, Salzmann S, Muller CS, Ekanayake-Mudiyanselage S, Elsner P, Thiele JJ (2002) Photoaging is associated with protein oxidation in human skin in vivo. J Invest Dermatol 118:618–625

37. Scharffetter-Kochanek K, Brenneisen P, Wenk J, et al (2000) Photoaging of the skin: From phenotype to mechanisms. Exp Gerontol 35:307–316

38. Smith JG, Davidson EA, Sams WM, Clark RD (1962) Alterations in human dermal connective tissue with age and chronic sun damage. J Invest Dermatol 39:347–350

39. Stewart SA, Ben-Porath I, Carey VJ, O'Connor BF, Hahn WC, Weinberg RA (2003) Erosion of the telomeric single-strand overhang at replicative senescence. Nat Genet 33:492–6

40. Talwar HS, Griffiths CE, Fisher GJ, Hamilton TA, Voorhees JJ (1995) Reduced type I and type III procollagens in photodamaged adult human skin. J Invest Dermatol 105:285–29041

41. Trifunovicz A, Wredenberg A, Falkenberg M, Spelbrink JN, Rovio AT, Bruder E, Bohlooly YM, Gidlof S, Oldfors A, Wibom R, Tornell J, Jacvobs HT, Lrsson NG (2004) Premature ageing in mice expressing defective mitochondrial DNA polymerase. Nature 429:417–42342

42. van Steensel B, Smogorzewska A, de Lange T (1998) TRF2 protects human telomeres from end-to-end fusions. Cell 92:401–13

43. Varani J, Schuger L, Dame MK, Leonhard Ch, Fligiel SEG, Kang S, Fisher GJ, Vorhees JJ (2004) Reduced fibroblast interaction with intact collagen as a mechanism for depressed collagen synthesis in photodamaged skin. J Invest Dermatol 122:1471–1479

44. Vijg J (2000) Somatic mutations and aging: a reevaluation. Mutat Res 447:117–135

45. von Zglinicki T (2000) Role of oxidative stress in telomere length regulation and replicative senescence. Ann N Y Acad Sci 908:99–110

46. von Zglinicki T, Saretzki G, Ladhoff J, d'Adda di Fagagna F, Jackson SP (2000) Human cell senescence as a DNA damage response. Mech Ageing Dev 126:111–7

47. Wallace DC (1992) Mitochondrial genetics: a paradigm for aging and degenerative diseases? Science 256:628–632

48. Wenk J, Brenneisen P, Meewes C, Wlaschek M, Peters T, Blaudschun R, Ma W, Kuhr L, Schneider L, Scharffetter-Kochanek K (2001) UV-induced oxidative stress and photoaging. Curr Probl Dermatol 29:83–94

49. Yang JH, Lee HC, Wei YH (1995) Photoageing-associated mitochondrial DNA length mutations in human skin. Arch Dermatol Res 287:641–8

50. Yano K, Ouira H, Detmar M (2002) Targeted over expression of the angiogenesis inhibitor thrombospondin-1 in the epidermis of transgenic mice prevents ultraviolet-B-induced angiogenesis and cutaneous photodamage. J Invest Dermatol 118:800–805

Chapter 5

Premature Skin Aging by Infrared Radiation, Tobacco Smoke and Ozone

5

Peter Schroeder, Stefan M. Schieke, Akimichi Morita

Contents

5.1 Introduction

Extrinsic skin aging has for many years been mainly attributed to ultraviolet (UV) radiation. More recently it has become evident that other factors contribute as well. These inducers include infrared radiation (IR), tobacco smoke and ozone. In this chapter we summarize the current state of knowledge concerning the epidemiology, molecular principles and prevention/protection relating to skin aging induced by these environmental factors.

5.2 Infrared Radiation

5.2.1 Physical Basics

Solar radiation reaching the earth's surface includes wavelengths in the range 290–4,000 nm and is divided into three bands: UV radiation (290–400 nm), visible light (400–760 nm) and IR (760–4,000 nm). IR is further divided into IR-A (λ=760–1,440 nm), IR-B (λ=1,440–3,000 nm) and IR-C (λ=3,000 nm to 1 mm). The main source of IR is the sun, but artificial IR sources are constantly gaining importance. They are used for therapeutic as well as for lifestyle purposes, e.g. for "wellness" irradiations or for skin rejuvenation which appears to be quite paradoxical.

While the photon energy of IR is lower than that of UV, the total amount of energy transferred by the sun is approximately 54% IR while UV only accounts for 7% [27]. Most of the IR lies within the IR-A band (approximately 30% of total solar energy), which deeply penetrates human skin, while IR-B and IR-C only penetrate the upper skin layers (Fig. 5.1) [10, 21, 27].

5

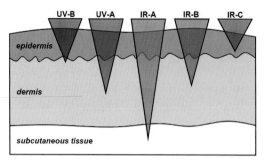

Fig. 5.1. Skin penetration of non-ionizing radiation. Different wavelengths of natural and artificial radiation have different penetration capabilities. In healthy skin, only very small amounts of UV-B penetrate past the epidermis. While only some UV-A reaches the dermis, approximately 50% of IR-A is absorbed in the dermis. IR-B reaches the dermis as well, while IR-C is completely absorbed in the epidermis

The actual solar dose reaching the skin is influenced by several factors: ozone layer, position of the sun, latitude, altitude, cloud cover and ground reflections.

5.2.2 Infrared Radiation and Skin Aging

The relevance of IR for premature skin aging was discussed by Kligman over 20 years ago [26]. She was the first to report that UV-induced skin damage in guinea pigs is enhanced by IR irradiation which prompted her to investigate the effect of IR alone. In her study, IR led to elastosis, with "IR inducing the production of many fine, feathery fibers" and "a large increase in ground substance, a finding also seen in actinically damaged human skin". From these observations she concluded that IR contributes to skin aging.

5.2.3 Molecular Mechanisms

It took 20 years until the first insight into the mechanism of IR-induced skin damage was obtained. Schieke et al. reported in 2002 [43] that low physiologically relevant doses of IR-A in vitro lead to an induction of matrix metalloproteinase-1 (MMP-1) in human dermal fibro-

blasts. Matrix metalloproteinases (MMPs) are zinc-dependent endopeptidases responsible for the degradation of extracellular matrix (ECM) components such as collagen and elastin. Under physiological conditions, MMPs are part of a coordinated network and are precisely regulated by their endogenous inhibitors, tissue inhibitors of MMPs (TIMPs). The unbalanced activity of MMPs with excessive proteolysis is thought to be a major pathophysiological factor in extrinsic skin aging. The increased expression of MMPs results in cleavage of fibrillar collagen, and thus impairs the structural integrity of the dermis [6, 16, 17, 42]. Recent in vivo investigations have confirmed that IR-A irradiation leads to an upregulation of MMP-1 expression in human skin. Irradiation with physiologically relevant doses of water-filtered IR-A causes an upregulation of MMP-1 at the mRNA and protein levels in irradiated skin areas compared to sham-irradiated skin from the same individuals [45, 46].

The IR-A-induced signaling pathway relevant for MMP-1 induction (Fig. 5.2) has been found to involve the activation of MAP kinases (MAPK). In general, three distinct MAPK pathways have been characterized: the extracellular signal-regulated kinase 1/2 (ERK1/2) pathway (Raf-MEK1/2-ERK1/2), and the c-Jun N-terminal kinase (MEKK1/3-MKK4/7-JNK1/2/3) and p38 (MEKK-MKK3/6-p38 a–d) pathways, also termed stress-activated protein kinases (SAPKs). The ERK1/2 pathway is primarily induced by mitogens such as growth factors, whereas the SAPK pathways are predominantly induced by inflammatory cytokines as well as environmental stress such as UV, heat and osmotic shock. Activated MAPKs translocate to the nucleus, where they phosphorylate and activate transcription factors such as c-Jun, c-Fos, ATF-2 and ternary complex factors (TCF) leading to the formation and activation of homo- or heterodimeric forms of the transcription factor AP-1 [9, 20, 29]. The promoter region of MMP-1 carries multiple AP-1-binding sites [1]. For IR-A, Schieke et al. have demonstrated that ERK1/2 and p38 are activated in dermal fibroblasts, and that inhibition of ERK1/2 activation subdues the IR-A-induced increase of MMP-1 [44].

Although the main research focus has been on MMP-1 it is very likely that the IR-A-induced

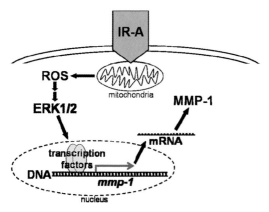

Fig. 5.2. IR-A-induced signal transduction (modified from Ref. 44). IR-A irradiation leads to an increase in the amount of mitochondrial ROS which in turn leads to activation of ERK1/2, finally resulting in an increased expression of MMP-1 mRNA and protein.

activation of MAPK affects the regulation of other genes as well.

5.2.4 Photobiology of IR-A

Reactive oxygen species (ROS) are established initial triggers of the molecular responses described above. Several studies have indicated that ROS cause an inactivation of protein-tyrosine phosphatases (PTPs) by oxidizing the conserved cysteine residue in the active sites of PTPs, thereby lead to a net increase in kinase phosphorylation/activation [4, 14, 35, 40]. In very recent studies ROS have been shown to be generated after IR-A irradiation and as well to mediate IR-A-induced upregulation of MMP-1 [45]. After IR-A irradiation the cellular ROS levels are increased in irradiated cells and a disturbance of the cellular glutathione (GSH) equilibrium occurs. GSH can prevent or repair oxidative damage, and as a consequence it is oxidized itself, forming the glutathione dimer (GSSG). In this regard, IR-A irradiation leads to a significant shift of the GSH/GSSG equilibrium towards the oxidized form [45].

IR-A-induced ROS production is of functional relevance because increasing the cellular GSH content abrogates the IR-A-induced upregulation of MMP-1 [45].

Future studies will have to identify the type and source of ROS involved in the IR-A response. It is likely that the photobiological mechanisms involved in IR-A-induced signal transduction differ from those observed in UV-induced signaling, since different chromophores are involved [25]. While the endogenous chromophores for IR remain to be identified, exogenous chromophores for IR, which are used for therapeutic purposes, e.g. in photodynamic therapy [28, 51], include palladium-bacteriopheophorbide and indocyanine green.

5.2.5 Protection Against IR-A

Up to now photoprotection of human skin has been limited specifically to UV radiation. The studies discussed above indicate, however, that protection against IR-A radiation has to be included in sunscreens in order to achieve optimal protection.

5.3 Tobacco Smoke

5.3.1 Epidemiological Studies

In 1971, Daniell reported that smoking has deleterious effects on the skin and that wrinkles are typical clinical features of smokers [12]. A recent epidemiological study provides evidence that tobacco smoking is one of the most important factors contributing to premature skin aging. Using silicone rubber replicas combined with computerized image processing, the association between wrinkle formation and tobacco smoking has been investigated [62]. The variance and depth of furrows in subjects with a history of smoking 35 pack-years or more were significantly greater than in non-smokers [58, 59, 62].

In another cross-sectional study, sun exposure, pack-years of smoking history and potential confounding variables were assessed by questionnaire. Facial wrinkles were quantified using the Daniell score. Logistic statistic analysis of the data revealed that age (OR=7.5), pack-years (OR=5.8) and sun exposure (OR=2.65) independently contributed to the formation of

5

facial wrinkles [61]. Similarly, tobacco smoking alone, independently of UV exposure, is strongly correlated with skin aging [31].

5.3.2 Molecular Mechanisms of Tobacco Smoke-Induced Skin Aging

Tobacco smoke contains numerous compounds; at least 3,800 constituents are known [2]. Which constituent(s) damage connective tissue of human skin and thereby cause wrinkles is still unclear. Tobacco smoking probably exerts its deleterious effects on skin through two major pathomechanisms: (1) directly by affecting epidermal integrity and (2) indirectly via dermal vessels [19, 32]. Smoking can decrease the moisture content of the facial stratum corneum (SC) which in turn contributes to facial wrinkling. Pursing the lips during smoking with contraction of facial muscles as well as squinting because of the irritation of smoke may enhance the formation of wrinkles.

Molecular mechanisms of tobacco smoke-induced damage are still poorly understood. In general tobacco smoke acts via mechanisms similar to those identified for UV radiation [15, 47, 60]. These include (1) a decrease in collagen and (2) an increase in tropoelastin content of the skin, (3) alterations in proteoglycan expression, and (4) induction of MMPs.

There is indeed evidence that the biosynthesis of new collagen fibers is significantly decreased by tobacco smoke extracts in cultured skin fibroblasts [60]. These studies have also shown that the production of procollagen types I and III, the precursors of collagen, is significantly reduced in supernatants of cultured fibroblasts that have been treated with tobacco smoke extracts [47].

Although elastic fibers account for only 2–4% of the ECM, they are critical for elasticity and resilience of normal skin. Accumulation of abnormal elastic material (i.e. solar elastosis) is a prominent histopathological feature of photoaged skin [36, 52]. Boyd et al. have reported that tobacco smoking (average of 42 pack-years of tobacco smoking) enhances elastosis [5]. Also, in an in vitro study, tobacco smoke extracts

elevated the expression of tropoelastin mRNA in cultured skin fibroblasts. Tropoelastin is the soluble precursor of elastin-forming elastic fibers, and this effect might thus be relevant for tobacco smoke-induced elastosis.

Besides collagen and elastin, proteoglycans are important ECM components. Versican, a large chondroitin sulfate (CS) proteoglycan, is present in the dermis in association with elastic fibers which contain a hyaluronic acid-binding domain. The core protein is thought to facilitate the binding of these macromolecules to other matrix components or cytokines such as transforming growth factor (TGF) [18]. Decorin, a small CS proteoglycan, has been shown to be co-distributed with collagen fibers and probably plays a role in cell recognition by connecting ECM components and cell surface glycoproteins [65]. Targeted disruption of decorin synthesis in mice results in a significant reduction in the tensile strength of skin [13]. It is therefore of interest that tobacco smoke extracts decrease versican levels in cultured skin fibroblasts, while increasing decorin content [63]. The in vivo relevance of these findings is currently under investigation. It has been observed in vivo that aged skin is characterized by a decrease in large CS proteoglycan (versican) and a concomitant increase in the proportion of small dermatan sulfate proteoglycan (decorin) [8]. Ito et al. have shown that versican is abundant in young rats and stains faintly in old rats, while decorin stains faintly in young animals [22]. There have been several reports on changes of proteoglycans in photoaged skin [3, 33]. Analysis of proteoglycan de novo synthesis has shown a marked increase after UVB exposure in mice [33]. Versican and decorin immunostaining have been found to be increased in tissue samples from photoaged skin [3]. It remains to be determined whether similar changes can also be induced by tobacco smoke.

Besides the synthesis of ECM components their breakdown is of crucial importance for ECM composition. The mRNA of the ECM-associated proteases MMP-1 and MMP-3 is induced in a dose-dependent manner in cultured skin fibroblasts stimulated with tobacco smoke extracts while the expression of the respective tissue inhibitors TIMP-1 and TIMP-3 remains unchanged [60]. In vitro studies indicate that tobacco smoking can cause an increased proteolysis of collagen fibers. This concept has been confirmed in a clinical study, which demonstrated significantly higher levels of MMP-1 mRNA in the buttock skin of smokers compared to non-smokers [30]. MMPs comprise a family of degradative enzymes which are not only responsible for the break down of collagen but also of other ECM components such as various proteoglycans. Accordingly MMP-7, a protease involved in proteoglycan degradation, is induced by tobacco smoke extracts in cultured fibroblasts [41].

Similar to UV- and IR-A-induced MMP expression, that induced by tobacco smoke appears to be mediated by ROS. NaN_3, L-ascorbic acid and vitamin E, which are all well-established quenchers of singlet oxygen, abrogate the induction of MMPs after exposure of fibroblasts to tobacco smoke extract. Among these antioxidants, L-ascorbic acid is the most effective [60].

TGF beta (TGF-β1) is a multifunctional cytokine that regulates cell proliferation and differentiation, tissue remodeling, and repair [34]. TGF-β1 is a potent inhibitor of epidermal growth, whereas in the dermis, TGF-β1 acts as a positive growth factor by inducing the synthesis of ECM proteins. TGF-β1 signals through a heteromeric complex of the type I/II TGF-β1 receptors, which ultimately initiates signal transduction [23, 37]. A recent study has shown that UV radiation causes downregulation of TGF-β type II receptor mRNA and protein expression as well as induction of Smad7 mRNA and protein expression in human skin [39].

In this regard tobacco smoke extracts have been found to induce the latent form, but not the active form, of TGF-β in vitro in cultured skin fibroblasts [64]. Because fibroblast responses to TGF-β1 are mediated through its active form, it has been proposed that tobacco smoke extracts may inhibit cellular responsiveness to TGF-β1 through the induction of the non-functional latent form and downregulation of TGF-β1 receptor [64].

In summary, similar to UV and IR-A radiation, tobacco smoke causes extrinsic skin aging. The underlying mechanisms are poorly understood and in certain aspects may overlap, and in other aspects may be different from, those induced by UV. More studies are clearly needed because precise knowledge of the mechanistic basis of tobacco smoke-induced skin aging is an indispensable prerequisite for the development of effective protective measures. Nonetheless, virtually all tobacco smoke-induced changes in skin could readily be avoided by not smoking.

5.4 Ozone and Skin Aging

Ozone (O_3) is a highly reactive oxidant that occurs naturally in the Earth's upper atmosphere (stratosphere) where it functions as a shield against harmful solar UV radiation. Depletion of stratospheric ozone by chlorofluorocarbons (CFCs) and other industrially produced ozone-depleting substances presents a major health problem. However, this protective stratospheric compound becomes a noxious pollutant with detrimental effects on human health, plants and ecosystems when present at ground level (troposphere). Ozone is one of the dominant components of photochemical smog and is a major air quality problem in urban as well as rural areas. Epidemiological studies have shown that exposure to ozone increases the morbidity and mortality from cardiovascular and respiratory diseases [7].

Photochemical smog develops from the reaction of air pollutants such as nitrogen oxides (NO_x) and volatile organic compounds (VOC) created from fossil fuel combustion in the presence of sunlight (hν) (Fig. 5.3). Emissions from industrial facilities and motor vehicle exhausts are the primary sources of NO_x and VOC. The specific climatic conditions (sunny and warm weather) which favor the development of photochemical smog/ozone make it a major health concern during the summer.

The skin as the organism's outermost barrier and the respiratory tract are the organs primarily affected by exposure to ozone. While several epidemiological as well as molecular studies have examined in detail the effects of ozone on the respiratory system, there are only a small number of studies that have investigated its effect on cutaneous tissues. However, these studies provide ample evidence that ozone as a

5

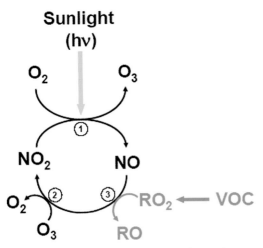

Fig. 5.3. Production of ozone and photochemical smog. Nitric oxide (NO) produced by the photodecomposition of nitrogen dioxide (NO_2) by sunlight reacts with molecular oxygen to give ozone (O_3) (*1*) being consumed again by NO (*2*) which happens naturally in an unpolluted atmosphere. The presence of volatile organic compounds (*VOC*) results in an alternative reaction pathway for NO not consuming the ozone (*3*), the concentrations of which can be elevated to toxic levels (i.e. photochemical smog)

strong oxidative agent is capable of affecting the integrity of skin.

The major targets of ozone in the skin are the superficial epidermal layers, and of these the SC is in direct contact with the environment.

Experimental evidence shows that ozone can induce oxidative stress and damage in the epidermis of murine skin. Exposure of mice to ozone results in depletion of antioxidants such as α-tocopherol (vitamin E) and ascorbic acid (vitamin C) in the superficial epidermal layers. In addition, the lipid peroxidation product malondialdehyde (MDA) can be detected in upper as well as lower epidermal layers [48].

These initial in vivo results have been supported by those of other studies on ozone toxicology demonstrating the depletion of lipophilic as well as hydrophilic antioxidants in murine epidermal skin. The same authors also demonstrated that there is also oxidative damage to lipids and proteins as measured by MDA and protein carbonyl levels as products of lipid and protein oxidation, respectively [49, 57]. Interestingly, concomitant exposure to ozone enhances the depletion of vitamin E induced by UV irradiation in hairless mice [53].

In addition to these biochemical markers of oxidative stress induced by ozone in superficial epidermal layers, it has recently been shown that ozone is capable of eliciting stress response mechanisms in deeper skin layers [54–56]. The expression of stress genes such as heat shock proteins (Hsp) 27 and 70 and heme oxygenase-1 (HO-1), specific stress proteins activated by several endogenous and exogenous stimuli, is induced in murine skin by ozone exposure. Furthermore, increased levels of cyclooxygenase-2 (COX-2) and inducible nitric oxide synthase (iNOS), and activation of the NFκB pathway, in ozone-exposed skin indicates a potential proinflammatory effect of ozone.

Interestingly, the expression of MMP-9 has also been found to be increased in response to ozone stimulation which points to a possible role of ozone in ECM remodeling, a prominent hallmark of photoaged skin.

These results indicate that, in addition to oxidatively damaging superficial layers, ozone also affects gene expression in deeper layers eliciting specific stress response mechanisms similar to the cutaneous response to UV radiation.

The superficial localization of ozone-induced oxidative damage to skin should make it accessible to preventive supplementation of antioxidants. In fact, Thiele et al. [50] have shown that topically applied vitamin E decreases MDA levels and thus lipid peroxidation in ozone-exposed murine skin. Cotovio et al. [11] used human keratinocytes and reconstructed human epidermis to show that vitamin C, N-acetylcysteine, an antioxidative thiol compound and antioxidative plant extract (green tea polyphenols) significantly decrease the formation of intracellular ROS in response to ozone.

The studies described above have convincingly shown that ozone is capable of affecting the integrity of the skin, with the SC as the major target for ozone-induced oxidative stress and depletion of antioxidants. Based on its high reactivity with abundant lipids in the SC, ozone is unlikely to penetrate much more deeply into the skin. Under these circumstances, how is ozone able to trigger gene expression and induction of stress proteins in deeper layers of the skin?

Pryor et al. [38] suggested a cascade mechanism by which ozone affects deeper tissue layers via the production of second messengers. Lipid peroxidation in superficial tissue layers at the air/organ interface results in the production of lipid ozonation products (LOP) which are small, diffusible molecules penetrating the superficial layers and triggering stress response and inflammatory mechanisms [24]. Although this concept was originally suggested to explain ozone-induced effects in the respiratory tract, it is also an attractive model for the cutaneous effects of ozone. However, further studies are required to confirm this mechanism in cutaneous tissues.

Despite the emerging knowledge on the cutaneous molecular effects of ozone exposure, no direct cause–effect relationship for skin aging has been described so far. However, the effects elicited upon exposure to ozone ranging from oxidative damage with depletion of antioxidants in the SC to activation of stress response mechanisms in deeper layers could contribute to a variety of skin pathologies inducing premature aging of the skin. In particular, it is conceivable that chronic exposure to ozone as a constituent of photochemical smog, by impairing the antioxidative capacity of the skin, aggravates the cutaneous damage induced by concomitant UV irradiation. With the increasing relevance of ozone as an air pollutant for humans, further studies are needed which specifically address the potential of ozone to induce skin aging and to establish possible preventive measures.

5.5 Concluding Remarks

As skin physiology and skin aging are very complex processes, it is not surprising that ongoing research efforts uncover more and more noxious substances with detrimental effects on the skin. Expanding the field of view, the effects of IR radiation, tobacco smoke, and ozone have received increasing attention. Besides the understanding of the pathomechanisms involved, questions of adequate protection arise, challenging modern research to counter premature skin aging. Life-style is an important topic as tobacco smoke and ozone can be avoided in

our modern society, and IR-A can be avoided at least partially. Nonetheless, the major portion of our daily IR-A load cannot be avoided and therefore additional protective approaches must be investigated.

References

1. Angel P, Baumann I, Stein B, Delius H, Rahmsdorf HJ, Herrlich P (1987) 12-O-tetradecanoyl-phorbol-13-acetate induction of the human collagenase gene is mediated by an inducible enhancer element located in the 50-flanking region. Mol Cell Biol 7:2256–2266
2. Bartsch H, Malaveille C, Friesen M, Kadlubar FF, Vineis P (1993) Black (air-cured) and blond (flue-cured) tobacco cancer risk. IV: Molecular dosimetry studies implicate aromatic amines as bladder carcinogens. Eur J Cancer 29A:1199–1207
3. Bernstein EF, Fisher LW, Li K, LeBaron RG, Tan EM, Uitto J (1995) Differential expression of the versican and decorin genes in photoaged and sun-protected skin: comparison by immunohistochemical and northern analyses. Lab Invest 72:662–669
4. Blanchetot C, Tertoolen LG, den Hertog J (2002) Regulation of receptor protein-tyrosine phosphatase alpha by oxidative stress. EMBO J 21:493–503
5. Boyd AS, Stasko T, King LE Jr, Cameron GS, Pearse AD, Gaskell SA (1999) Cigarette smoking-associated elastotic changes in the skin. J Am Acad Dermatol 41:23–26
6. Brenneisen P, Sies H, Scharffetter-Kochanek K (2002) Ultraviolet-B irradiation and matrix metalloproteinases: from induction via signaling to initial events. Ann N Y Acad Sci 973:31–43
7. Brunekreef B, Holgate ST (2002) Air pollution and health. Lancet 360:1233–1242
8. Carrino DA, Sorrell JM, Caplan AI (2000) Age-related changes in the proteoglycans of human skin. Arch Biochem Biophys 373:91–101
9. Chang LF, Karin M (2001) Mammalian MAP kinase signaling cascades. Nature 410:37–40
10. Cobarg CC (1995) Physikalische Grundlagen der wassergefilterten Infrarot-A-Strahlung. In: Vaupel P, Krüger W (eds) Wärmetherapie mit wassergefilterter Infrarot-A-Strahlung. Hippokrates Verlag, Stuttgart, p 19
11. Cotovio J, Onno L, Justine P, Lamure S, Catroux P (2001) Generation of oxidative stress in human cutaneous models following in vitro ozone exposure. Toxicol In Vitro 15:357–362
12. Daniell HW (1971) Smoker's wrinkles. A study in the epidemiology of "crow's feet". Ann Intern Med 75:873–880

5

13. Danielson KG, Baribault H, Homes DF, Graham H, Kadler KE, Iozzo RV (1997) Targeted disruption of decorin leads to abnormal collagen fibril morphology and skin fragility. J Cell Biol 136:729–743
14. Finkel T (2003) Oxidant signals and oxidative stress. Curr Opin Cell Biol 15:247–254
15. Fisher GJ, Voorhees JJ (1998) Molecular mechanisms of photoageing and its prevention by retinoic acid: ultraviolet irradiation induces MAP kinase signal transduction cascades that induce Ap-1-regulated matrix metalloproteinases that degrade human skin in vivo. J Investig Dermatol Symp Proc 3:61–68
16. Fisher GJ, Wang ZQ, Datta SC, Varani J, Kang S, Voorhees JJ (1997) Pathophysiology of premature skin aging induced by ultraviolet light. N Engl J Med 337:1419–1428
17. Fisher GJ, Kang S, Varani J, Bata-Csorgo Z, Wan Y, Datta S, Voorhees JJ (2002) Mechanisms of photoaging and chronological skin aging. Arch Dermatol 138:1462–1470
18. Fisher LW, Termine JD, Young MF (1989) Deduced protein sequence of bone small proteoglycan I (biglycan) shows homology with proteoglycan II (decorin) and several nonconnective tissue proteins in a variety of species. J Biol Chem 264:4571–4576
19. Frances C (1998) Smoker's wrinkles: epidemiological and pathogenic considerations. Clin Dermatol 16:565–570
20. Hazzalin CA, Mahadevan LC (2002) MAPK-regulated transcription: a continuously variable gene switch? Nat Rev Mol Cell Biol 3:30–40
21. Hellige G, Becker G, Hahn G (1995) Temperaturverteilung und Eindringtiefe wassergefilterter Infrarot-A-Strahlung. In: Vaupel P, Becker G (eds) Wärmetherapie mit wassergefilterter Infrarot-A-Strahlung. Hippokrates Verlag, Stuttgart, p 63
22. Ito Y, Takeuchi J, Yamamoto K, Hashizume Y, Sato T, Tauchi H (2001) Age differences in immunohistochemical localizations of large proteoglycan, PG-M/versican, and small proteoglycan, decorin, in the dermis of rats. Exp Anim 50:159–166
23. Kadin ME, Cavaille-Coll MW, Gertz R, Massague J, Cheifetz S, George D (1994) Loss of receptors for transforming growth factor beta in human T-cell malignancies. Proc Natl Acad Sci U S A 91:6002–6006
24. Kafoury RM, Pryor WA, Squadrito GL, Salgo MG, Zou X, Friedman M (1999) Induction of inflammatory mediators in human airway epithelial cells by lipid ozonation products. Am J Respir Crit Care Med 160:1934–1942
25. Karu T (1999) Primary and secondary mechanisms of action of visible to near-IR radiation on cells. J Photochem Photobiol B 49:1
26. Kligman LH (1982) Intensification of ultraviolet-induced dermal damage by infrared radiation. Arch Dermatol Res 272:229
27. Kochevar IE, Pathak MA, Parrish JA (1999) Photophysics, photochemistry, and photobiology. In: Freedberg IM, Eisen AZ, Wolff K, Austen KF, Goldsmith LA, Katz SI, Fitzpatrick TB (eds) Fitzpatrick's dermatology in general medicine. McGraw-Hill, New York, p 220
28. Koudinova NV, Pinthus JH, Brandis A, Brenner O, Bendel P, Ramon J, Eshhar Z, Scherz A, Salomon Y (2003) Photodynamic therapy with Pd-bacteriopheophorbide (TOOKAD): successful in vivo treatment of human prostatic small cell carcinoma xenografts. Int J Cancer 104:782
29. Kyriakis JM, Avruch J (2001) Mammalian mitogen-activated protein kinase signal transduction pathways activated by stress and inflammation. Physiol Rev 81:807–869
30. Lahmann C, Bergemann J, Harrison G, Young AR (2001) Matrix metalloprotease-1 and skin ageing in smokers. Lancet 357:935–936
31. Leung W-C, Harvey I (2002) Is skin ageing in the elderly caused by sun exposure or smoking? Br J Dermatol 147:1187–1191
32. Lofroth G (1989) Environmental tobacco smoke: overview of chemical composition and genotoxic components. Mutat Res 222:73–80
33. Margelin D, Fourtanier A, Thevenin T, Medaisko C, Breton M, Picard J (1993) Alterations of proteoglycans in ultraviolet-irradiated skin. Photochem Photobiol 58:211–218
34. Massague J (1998) TGF-beta signal transduction. Annu Rev Biochem 67:753–791
35. Meng TC, Fukada T, Tonks NK (2002) Reversible oxidation and inactivation of protein tyrosine phosphatases in vivo. Mol Cell 9:387–399
36. Montagna W, Kirchner S, Carlisle K (1989) Histology of sun-damaged human skin. J Am Acad Dermatol 21:907–918
37. Piek E, Heldin CH, Ten Dijke P (1999) Specificity, diversity, and regulation in TGF-beta superfamily signaling. FASEB J 13:2105–2124
38. Pryor WA, Squadrito GL, Friedman M (1995) The cascade mechanism to explain ozone toxicity: the role of lipid ozonation products. Free Radic Biol Med 19:935–941
39. Quan T, He T, Voorhees JJ, Fisher GJ (2001) Ultraviolet irradiation blocks cellular responses to transforming growth factor-beta by down-regulating its type-II receptor and inducing Smad7. J Biol Chem 276:26349–26356
40. Rao RK, Clayton LW (2002) Regulation of protein phosphatase 2A by hydrogen peroxide and glutathionylation. Biochem Biophys Res Commun 293:610–616

41. Saarialho-Kere U, Kerkela E, Jeskanen L, et al (1999) Accumulation of matrilysin (MMP-7) and macrophage metalloelastase (MMP-12) in actinic damage. J Invest Dermatol 113:664–672

42. Scharffetter K, Wlaschek M, Hogg A, Bolsen K, Schothorst A, Goerz G, Krieg T, Plewig G (1991) UVA irradiation induces collagenase in human dermal fibroblasts in vitro and in vivo. Arch Dermatol Res 283:506–511

43. Schieke S, Stege H, Kurten V, Grether-Beck S, Sies H, Krutmann J (2002) Infrared-A radiation-induced matrix metalloproteinase 1 expression is mediated through extracellular signal-regulated kinase 1/2 activation in human dermal fibroblasts. J Invest Dermatol 119:1323

44. Schieke SM, Schroeder P, Krutmann J (2003) Cutaneous effects of infrared radiation: from clinical observations to molecular response mechanisms. Photodermatol Photoimmunol Photomed 19:228

55. Schroeder P, Wild S, Schieke SM, Krutmann J (2004) Further analysis of infrared A (IRA) radiation-induced MMP-1 expression. J Invest Dermatol 122/3:A140

46. Schroeder P, Wild S, Marks C, Kuerten V, Krutmann J (2005) In vivo relevance of infrared A radiation induced matrix metalloproteinase-1 expression: differential effects on epidermis and dermis. Arch Dermatol Res 296 (in press)

47. Shuster S (2001) Smoking and wrinkling of the skin. Lancet 358:330

48. Thiele JJ, Traber MG, Tsang K, Cross CE, Packer L (1997a) In vivo exposure to ozone depletes vitamins C and E and induces lipid peroxidation in epidermal layers of murine skin. Free Radic Biol Med 23:385–391

49. Thiele JJ, Traber MG, Polefka TG, Cross CE, Packer L (1997b) Ozone-exposure depletes vitamin E and induces lipid peroxidation in murine stratum corneum. J Invest Dermatol 108:753–757

50. Thiele JJ, Traber MG, Podda M, Tsang K, Cross CE, Packer L (1997) Ozone depletes tocopherols and tocotrienols topically applied to murine skin. FEBS Lett 401:167–170

51. Tseng WW, Saxton RE, Deganutti A, Liu CD (2003) Infrared laser activation of indocyanine green inhibits growth in human pancreatic cancer. Pancreas 27:e42

52. Tsuji T (1987) Ultrastructure of deep wrinkles in the elderly. J Cutan Pathol 14:158–64

53. Valacchi G, Weber SU, Luu C, Cross CE, Packer L (2000) Ozone potentiates vitamin E depletion by ultraviolet radiation in the murine stratum corneum. FEBS Lett 466:165–168

54. Valacchi G, van der Vliet A, Schock BC, Okamoto T, Obermuller-Jevic U, Cross CE, Packer L (2002) Ozone exposure activates oxidative stress responses in murine skin. Toxicology 179:163–170

55. Valacchi G, Pagnin E, Okamoto T, Corbacho AM, Olano E, Davis PA, van der Vliet A, Packer L, Cross CE (2003) Induction of stress proteins and MMP-9 by 0.8 ppm of ozone in murine skin. Biochem Biophys Res Commun 305:741–746

56. Valacchi G, Pagnin E, Corbacho AM, Olano E, Davis PA, Packer L, Cross CE (2004) In vivo ozone exposure induces antioxidant/stress-related responses in murine lung and skin. Free Radic Biol Med 36:673–681

57. Weber SU, Thiele JJ, Cross CE, Packer L (1999) Vitamin C, uric acid, and glutathione gradients in murine stratum corneum and their susceptibility to ozone exposure. J Invest Dermatol 113:1128–1132

58. Wenk J, Brenneisen P, Meewes C, et al (2001) UV-induced oxidative stress and photoageing. Curr Probl Dermatol 29:83–94

59. Yin L, Morita A, Tsuji T (2000) Alterations of extracellular matrix induced by tobacco smoke extract. Arch Dermatol Res 292:188–194

60. Yin L, Morita A, Tsuji T (2000) Tobacco smoking: a role of premature skin ageing. Nagoya Med J 43:165–171

61. Yin L, Morita A, Tsuji T (2001) Skin ageing induced by ultraviolet exposure and tobacco smoking: evidence from epidemiological and molecular studies. Photodermatol Photoimmunol Photomed 17:178–183

62. Yin L, Morita A, Tsuji T (2001) Skin premature ageing induced by tobacco smoking: the objective evidence of skin replica analysis. J Dermatol Sci 27 [Suppl 1]:S26–31

63. Yin L, Morita A, Tsuji T (2002) Molecular alterations of tropoelastin and proteoglycans induced by ultraviolet A and tobacco smoke extracts in cultured skin fibroblasts. Nagoya Med J 45:63–74

64. Yin L, Morita A, Tsuji T (2003) Tobacco smoke extract induces age-related changes due to the modulation of TGF-b. Exp Dermatol 12:51–56

65. Zimmermann DR, Ruoslahti E (1989) Multiple domains of the large fibroblast proteoglycan, versican. EMBO J 8:2975–2981

Chapter 6

The Role of Hormones in Intrinsic Aging

6

Christos C. Zouboulis, Evgenia Makrantonaki

Contents

6.1 Introduction

Youth has been an adored sign of power and health since the appearance of mankind. Eternal youth was represented in Greek mythology by the goddess of beauty Aphrodite, in the ancient Roman world by her counterpart Venus, and in northern-Germanic mythology by the goddess Iduna; all showed remarkable similarities in attributed appearance and behavior. The painter Lucas Cranach the Older presents in his famous picture "The Youth Spring" (1546) how he and his contemporaries imagined rejuvenation.

Today, aging is a pressing medical and socioeconomic problem. A demographic revolution is underway throughout the world, with Europe, North America, and Japan in the front line. Over the past few years, the world's population has continued its remarkable transition from a state of high birth and death rates to a state of low birth and death rates. The major effect of this transition has been the growth in the number and proportion of older persons. Such a rapid, large and ubiquitous growth has never been seen in the history of civilization.

The current demographic revolution is predicted to continue well into the coming decades. Its major features include the following (source: Population Division, Department of Economic and Social Affairs, United Nations Secretariat):

■ Increase of average lifespan: over the last half of the 20th century, 20 years were added to the average lifespan, bringing global life expectancy to its current level of 66 years.

■ Increasing numbers of aged individuals: one out of every ten persons is now aged 60 years or above; by 2050, one out of five, and by 2150, one out of three persons will be aged 60 years or older.

■ The older population itself is aging: the oldest old (aged 80 years or older) is the fastest growing segment of the population.

6

On the other hand, health systems are urgently required to respond to the resulting epidemiological shift already observed in a number of developing countries. Since health systems are presently based on providing care to acute episodic conditions, they are not geared towards chronic care needs and especially care for the aged.

6.2 Hormones and Human Skin

Human skin is classically regarded as the target for several hormones whose effects have long been recognized and in some instances well characterized [1]. For example, hair follicles and sebaceous glands are the targets for androgen steroids secreted by the gonads and the adrenal cortex [2, 3] and melanocytes are directly influenced by polypeptide hormones of the pituitary [4]. In addition, hormones play an important role in the development and physiological function of human skin tissues [5, 6]. From the modern dermatoendocrinological point of view the skin is not only the recipient of signals from distant transmitters but is also an organized community in which the cells and organelles emit, receive and coordinate molecular signals from a seemingly unlimited number of distant sources, their neighbors, and themselves [7]. In the widest sense, the human skin and its cells are the targets as well as the producers of hormones. For example, the circulating androgens dehydroepiandrosterone (DHEA) and androstenedione are converted in the skin to testosterone and further to the more potent androgen 5α-dihydrotestosterone (5α-DHT) [3, 8].

6.3 Hormones and Skin Aging

Hormones are decisively involved in intrinsic aging which is accompanied by reduced secretion of the pituitary and adrenal glands, and the gonads [1, 9, 10], leading to characteristic body and skin phenotypes as well as behavior pattern. There are decreases in sexual desire, intellectual activity, lean body mass, erectile capacity in men, body hair and bone mineral density that results in osteoporosis, there are increases in fatigue, depression, visceral fat and obesity, and finally there are alterations in the skin. Women in the western world currently live one-third of their life (menopause) and men at least 20 years of their life (partial androgen deficiency of the aging male, PADAM) in a state of hormonal deficiency. However, our knowledge of whether there is a direct or indirect connection between hormonal deficiency and skin aging still remains limited. In females serum levels of 17β-estradiol, DHEA, progesterone, growth hormone (GH) and its downstream hormone insulin-like growth factor I (IGF-I) are significantly decreased with increasing age (Fig. 6.1) [11, 12]. Since progesterone controls androgens, a relative peripheral hyperandrogenism occurs in women with increasing age. In males, serum levels of GH and IGF-I significantly decrease, whereas 17β-estradiol only shows minor changes. Testosterone levels fall below the normal range in 20–30% of males aged 60–80 years [13]. Systemic substitution of estrogens in females and GH in males has been shown to lead to positive effects on several organs including signs of skin rejuvenation in both cases and to prevention of the further aging process in controlled studies [14, 15].

The first hormone whose serum levels change with age is melatonin. Its levels decrease as early as from the 5th up to the 20th year of life and stay at a low plateau until the age of 75 years. Serum levels of GH and DHEA decrease continuously from the 20th year of life until old age in males. Simultaneously, a decrease of testosterone and IGF-I occurs, which proceeds less steeply than the decrease of GH and DHEA. Estrogens in females decrease rapidly after the menopause and up to the 60th year of life and show a low level plateau thereafter [9].

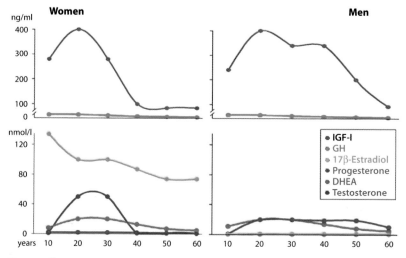

Fig. 6.1. Changes in serum concentrations of the most important hormones with increasing age in women and men

6.4 Estrogens

The major importance of estrogens for skin homeostasis and their effects during hormone replacement therapy are discussed in detail in Chapter 11. There is broad agreement that sex hormones are involved in the aging process, but it is also obvious that hormone replacement should not be administered as an independent treatment for skin aging [16, 17]. In addition to the data presented in Chapter 11, there is current evidence that 17β-estradiol increases the number of IGF-I receptors of basal, undifferentiated epidermal keratinocytes and, indirectly the number of IGF-I/IGF-I receptor complexes, whereas IGF-I is produced in fibroblasts [18, 19]. On the other hand, 17β-estradiol stimulates the synthesis of prostaglandin D2, which is metabolized to Δ12-Prostaglandin J2 [20]. Δ12-Prostaglandin J2 is one of the natural ligands of peroxisome proliferators activated receptor γ1, which is expressed in epithelial skin cells, including sebocytes, and stimulates sebaceous differentiation and lipogenesis [21]. A decline of 17β-estradiol, as occurs with the menopause, inhibits proliferation of epidermal keratinocytes and reduces sebaceous differentiation and lipogenesis. Interestingly, it has been shown that local application of 17β-estradiol stimulates se-

bum secretion on the face of post-menopausal women [14]. As in the first years after the menopause, compensation mechanisms are switched on by the sebaceous glands, so the hormonal substitution admittedly leads in this period of time to an improvement in skin quality with a reduction in the pore size. However, the rate of the sebum secretion is not being influenced [22].

Skin cells express estrogen receptors, making them directly susceptible to estrogens, but crosstalk between estrogen and IGF-I signaling pathways also appears to take place. IGF-I plays a major role in regulating lipid synthesis in sebocytes and proliferation in fibroblasts and may, therefore, modulate estrogen activity on normal and aged skin cells [23]. Other evidence suggests that at least phytoestrogens, such as genistein, regulate sebocyte differentiation through upregulation of PPARγ expression [24].

Lack of estrogens probably plays an important role in osteoporosis in men [25], but it is still unknown to what extent estrogens can influence skin quality and age-associated diseases in men [26]. Men with pathological glucose metabolism and elevated body mass have increased levels of estrogens in the serum [25]. The tissue levels of genistein, a phytoestrogen, negatively correlate with prostate mass, indicating a possible role in prevention or treatment of benign prostatic hypertrophy [27]. A prospective study, including 375 men aged 45–85 years, showed that estrogen levels correlate positively with biologically available testosterone and free testosterone in the serum [28]. On the other hand, this study, which examined a palette of hormone parameters, showed only a small interaction between the signs and symptoms of PADAM and serum estrogen levels.

6.5 Testosterone

With advancing age, men show a decrease in testosterone levels of up to 1.6% per year [29, 30]. Testosterone levels are below the bottom of the reference range of 12 nmol/l in 20% of men aged 60 years and in 35% of men aged over 80 years. Interestingly, recent data indicate that peripheral organs, and especially the skin, are capable of steroid biosynthesis, including synthesis of androgens [1, 3]. Thus, the sebaceous gland plays a decisive role by controlling steroid biosynthesis and androgen metabolism in the human skin. Testosterone treatment in old age has been connected with positive and negative effects [30]. The positive effects include increased physical strength, decreased fat tissue mass, enhanced bone density and improvement in libido and sexual function, especially in men with secondary hypogonadism [29]. Improvement in cognitive function has also been postulated. The negative effects include an increased rate of uncontrolled prostate malignancies, deterioration of existing sleep apnea syndrome, polyglobuly and gynecomastia. As far as cardiovascular function is concerned, both improvement as well as deterioration have been observed. The exact effects of testosterone deficiency on the skin are not known; however, recent animal research has shown a negative influence of testosterone on the epidermal barrier and wound healing [31, 32].

In women, androgen serum levels are reduced in individuals more than 9 years following the menopause [33]. There are some early hints that androgen treatment – up to the physiological serum concentration in females – can improve bone density, libido and quality of life [34, 35]. However, androgen treatment has also been associated with virilization in females after chronic systemic or topical application [22].

6.6 GH/IGF-I

It is most likely that GH and IGF-I assume the functions of testosterone in aging males. In a randomized study including 21 healthy men (61–81 years of age) with IGF-I serum levels below 350 E/l, 12 individuals were treated with 0.03 mg human GH over 6 months three times weekly, while 9 individuals served as controls [15]. The group receiving GH showed an 8.8% increase in muscle mass, a 14.4% reduction in fat tissue, a 1.6% improvement in bone density and a 7.1% increase in skin thickness. However, similar effects can be obtained with exercise [36], and elderly men often respond to GH administration with side effects such as carpal tunnel syndrome, peripheral edema, joint pain and swelling, gynecomastia, glucose intolerance, and possibly increased cancer risk [37, 38]. Low-dose GH treatment may bring advantages in certain patients [39]. Synergistic effects of GH and testosterone on protein metabolism and body composition in prepubertal boys [40] and, experimentally, on cortical bone formation and bone density of aged orchiectomized rats [41] have been observed. Nevertheless, further studies are required to identify the exact role of GH in aging individuals. Despite the decline of GH and IGF-I serum levels with increasing age no GH-related response of peripheral organs has been documented [42]. Under GH substitution in elderly individuals, an increase in muscle mass has only been confirmed in GH-deficient male patients [43]. In vitro studies indicate that testosterone, 5α-DHT and IGF-I stimulate proliferation and lipid synthesis in human sebocytes, whereas GH only stimulates lipid synthesis [44–46]. IGF-I also stimulates the proliferation of fibroblasts [46].

In order to investigate the influence of GH on aging parameters and its association with human longevity, IGF-I serum levels were examined in 250 healthy probands [47]. Elderly men (around the age of 70 years) with IGF-I serum levels similar to those of young men (up to the age of 39 years) showed no age-associated decrease in testosterone levels, no decrease in muscle mass and no increase in fat mass. These individuals exhibited no particular hormone metabolism characteristics or nutritional habits. Elderly men who showed a marked decrease of IGF-I serum levels died at a younger age than men of the same age whose IGF-I levels had not declined. Although alterations of IGF-I serum levels were not so prominent in female probands in this study, our own in vitro experiments suggest an indirect molecular effect of estrogens on the IGF-I/IGF-IR signaling pathway on the aging process of skin cells [46].

6.7 DHEA

It is still unclear whether the adrenal cortex hormone DHEA plays a role in the aging process of the organism. A decrease in DHEA serum levels and its precursor DHEA sulfate with increasing age may contribute to the known reduction in their metabolites, the sex hormones androgens and estrogens [1]. In vitro, DHEA influences the extracellular matrix. Inhibition of collagenase expression in human dermal fibroblasts at the mRNA level, stimulation of stromelysin-1 mRNA expression and a slight increase of 1(I)-procollagen mRNA levels have been documented [48]. In the first double-blind, placebo-controlled clinical study, oral administration of DHEA at 50 mg/day over 1 year in 280 healthy women and men aged 60–79 years led to a significant attenuation of the bone degradation activity and an increase in libido parameters in women older than 70 years. Skin quality was improved in the women ingesting DHEA with respect to hydration, epidermal thickness, sebum production and pigmentation. Circulating DHEA levels were similar to "young" levels and increases in circulating testosterone and 17β-estradiol without side effects were observed [49]. In contrast, no effect of DHEA administration was documented in healthy men after 1 year of DHEA replacement (50 mg/day) [50]. On the other hand, aging males with partial androgen deficiency experienced progressive improvement in mood, fatigue and joint pain during a 1 year of DHEA replacement at 25 mg/day [51].

In another prospective study including 375 men aged 45–85 years, negative correlations between DHEA sulfate serum levels with advancing age and sexual activity, general existential orientation and serum PSA levels were observed [52]. The results of this study, which examined diverse hormone parameters, showed a small interaction between the symptoms of PADAM and DHEA sulfate levels. In summary, current data do not indicate any consistent correlation between DHEA and aging parameters in healthy individuals.

6.8 Progesterone

C21 steroids, and especially progesterone, are involved in various physiological processes besides reproduction [53]. Progesterone is able to increase the number of Langerhans cells [54]. This may be a factor in the gender specificity of HIV1 infection, and also in the high premenstrual incidence of sexually transmitted diseases [55]. Progesterone suppresses the expression of some members of the matrix metalloproteinase superfamily and reduces collagenolytic activity [53]. These effects can be observed during pregnancy, but are not only apparent on the myometrium. A suppressive effect of progesterone on matrix metalloproteinases may also be observed in the skin, and this effect may be the reason why skin appears young [53]. Conversely, the decrease in progesterone with age may play a role in the development of the changes observed in aging skin.

6.9 Melatonin

Melatonin, a hormone produced in the glandula pinealis under the influence of β-adrenergic receptors, follows a circadian rhythm of secretion, with low serum levels during the day and an increase at night [56]. It influences the seasonal biorhythm, the sleep–awake rhythm and modulates immune-biological defense mechanisms. In addition, melatonin has an antioxidative potential [56, 57], and, as it is a lipophilic substance, possesses the ability to diffuse freely through cell membranes and to capture free radicals, especially hydroxyl radicals (·OH), in the extra- and intracellular matrix [57]. After oral administration, melatonin levels increase only minimally because of a high first-pass metabolism of over 90%. However, after topical application, the increase in plasma levels seems to be noticeable and long-lasting [58]. The topical application of a 0.5% melatonin nanocolloid gel seems to have a protective effect on the formation of UVB-induced erythema [59]. This effect could be useful for the prevention of light-induced skin aging. However, clinical studies with melatonin have not shown any positive effects on skin so far [56].

6.10 Cortisol, Thyroid Hormones and Vitamin D

In addition to the prominent endocrinological alterations with aging mentioned above, cortisol, thyroid hormones, and vitamin D may also be affected [60].

A recent clinical study has shown that in unfit women, aging is associated with greater hypothalamus pituitary adrenal axis reactivity to psychological stress with increases in heart rate, blood pressure, and circulating adrenocorticotropic hormone and cortisol [61]. However, this may not be a global aging-associated event, since neither fit women [61] nor DHEA-substituted aging men [51] exhibit increased cortisol levels.

Human epidermal keratinocytes in culture convert thyroxine to 3,5,3′-triiodothyronine by type II iodothyronine deiodination [62], and this function may be impaired with age. Some of the symptoms and signs of hypothyroidism and hyperthyroidism in elderly patients may be mistakenly attributed to "old age". Weight loss, muscle weakness, tremor, angina, congestive heart failure – all signs of hyperthyroidism – also occur with aging. Fatigue, sluggishness, withdrawal behavior, senile atrophic skin changes – all signs of hypothyroidism – are also a part of the normal aging process [63].

Osteoporosis is one major health condition that contributes to excess morbidity and mortality in women after the menopause [64]. Older adults require more vitamin D [65] and vitamin D insufficiency is a contributing cause of osteoporosis. Production of vitamin D in the skin is determined by length of exposure, which often decreases with aging, latitude, season, and degree of skin pigmentation. Black individuals produce less vitamin D_3 than white do in response to usual levels of sun exposure and have lower 25-hydroxyvitamin D [25(OH)D] concentrations in winter and summer [66]. Black Americans also use dietary supplements less frequently than white ones. However, blacks and whites appear to have similar capacities to absorb vitamin D and to produce vitamin D after repeated exposure to high doses of UVB light. There is a growing consensus on the basis of studies of older white subjects living in Europe and the United States [66], that serum 25(OH)D concentrations of at least 75–80 nmol/l are needed for optimal bone health. On the basis of studies conducted in the temperate zone, the intake of vitamin D_3 needed to maintain a group average 25(OH)D concentration of 80 nmol/l in winter is approximately 1,000 IU/day [66].

6.11 The Skin as Independent Peripheral Organ

The former perception that skin is only a target organ for hormones has been revised following recent studies. Skin has the ability to produce hormones by itself and fulfills all the criteria for an independent peripheral endocrine organ, indeed the largest endocrine organ (Fig. 6.2) [1, 67].

IGF-I, IGF-binding proteins, propiomelanocortin derivatives, catecholamines, steroids and vitamin D are formed in the skin by cholesterol, retinoids, carotenoids, eicosanoids and fatty acids. Hormones exert their biological effect on the skin through interaction with highly specific receptors. Skin metabolizes hormones and is capable of activating and inactivating them. These different activities take place normally in different cell populations of the skin and lead to an alignment of biological effects, which is important for skin homeostasis, as also for the entire organism. Such biological activities include the synthesis of vitamin D [65], sex hormones,

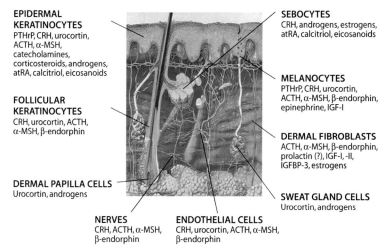

EPIDERMAL KERATINOCYTES
PTHrP, CRH, urocortin, ACTH, α-MSH, catecholamines, corticosteroids, androgens, atRA, calcitriol, eicosanoids

FOLLICULAR KERATINOCYTES
CRH, urocortin, ACTH, α-MSH, β-endorphin

DERMAL PAPILLA CELLS
Urocortin, androgens

NERVES
CRH, ACTH, α-MSH, β-endorphin

ENDOTHELIAL CELLS
CRH, urocortin, ACTH, α-MSH, β-endorphin

SEBOCYTES
CRH, androgens, estrogens, atRA, calcitriol, eicosanoids

MELANOCYTES
PTHrP, CRH, urocortin, ACTH, α-MSH, β-endorphin, epinephrine, IGF-I

DERMAL FIBROBLASTS
ACTH, α-MSH, β-endorphin, prolactin (?), IGF-I, -II, IGFBP-3, estrogens

SWEAT GLAND CELLS
Urocortin, androgens

Fig. 6.2. Synthesis of hormones in human skin. *PTHrP* parathyroid hormone-related peptide, *CRH* corticotrophin-releasing hormone, *ACTH* adrenocorticotropic hormone, *α-MSH* α-melanocyte-stimulating hormone, *atRA* all-*trans* retinoic acid, *IGF-I* insulin-like growth factor-I, *IGF-II* insulin-like growth factor-II, *IGFBP3* insulin-like growth factor-binding protein-3 (from Ref. 67)

especially after the menopause [8], and IGF-binding proteins [42].

Moreover, there is increasing evidence that human skin produces hormones which are released into the circulation and are important for the function of the entire human organism [68]. The abundance of IGF-binding protein 3 message is greater in the skin than in the liver and circulating IGF-binding protein 3 concentrations are significantly increased by GH and IGF-1 [69]. Interestingly, IGF-binding protein 3 has an innate strong inhibitory effect on the growth of epithelial tumors. GH has a direct effect on the regulation of IGF-binding protein 3 synthesis, and the response of skin IGF-binding protein 3 mRNA levels to both GH and IGF-I suggests that dermal fibroblasts could be more important than the liver in the regulation of the circulating reservoir IGF-binding protein 3 in certain circumstances.

Another example is sex steroids. Whereas a large proportion of androgens and estrogens in men and women are synthesized locally in peripheral target tissues from the inactive adrenal precursors DHEA and androstenedione [70], DHEA and androstenedione are converted to testosterone and further to 5α-DHT by the intracellular enzyme 5α-reductase in the periphery. Thus the skin is the source of considerable amounts of the circulating testosterone and 5α-DHT. Circulating testosterone is coproduced in the skin and in other peripheral organs. The best estimate of the intracrine formation of estrogens in peripheral tissues in women is in the order of 75% before the menopause and close to 100% after the menopause, except for a small contribution from ovarian and/or adrenal testosterone and androstenedione [7]. Thus, in postmenopausal women, almost all active sex steroids are made in target tissues by an intracrine mechanism.

The exact relevance of the endocrine function of the skin for its own aging process remains to be elucidated.

References

1. Zouboulis CC (2000) Human skin: an independent peripheral endocrine organ. Horm Res 54:230–242
2. Ebling FJG (1990) The hormonal control of hair growth. In: Orfanos CE, Happle R, (eds) Hair and hair growth. Springer, Berlin Heidelberg New York, pp 267–99
3. Fritsch M, Orfanos CE, Zouboulis CC (2001) Sebocytes are the key regulators of androgen homeostasis in human skin. J Invest Dermatol 116:793–800

6

4. Suzuki I, Cone CD, Im S, Nordlund J, Abdel-Malek ZA (1996) Binding of melanocortin hormones to the melanocortin receptor MC1R on human melanocytes stimulates proliferation and melanogenesis. Endocrinology 137:1627–1633

5. Deplewski D, Rosenfield RL (2000) Role of hormones in pilosebaceous unit development. Endocr Rev 21:363–392

6. Stenn KS, Paus R (2001) Control of hair follicle cycling. Physiol Rev 81:449–494

7. Labrie F, Luu-The V, Labrie C, Pelletier G, El-Alfy M (2000) Intracrinology and the skin. Horm Res 54:218–229

8. Orfanos CE, Adler YD, Zouboulis CC (2000) The SAHA syndrome. Horm Res 54:251–258

9. Phillips TJ, Demircay Z, Sahu M (2001) Hormonal effects on skin aging. Clin Geriatr Med 17:661–672

10. Thiboutot DM (1995) Dermatological manifestations of endocrine disorders. J Clin Endocrinol Metab 80:3082–3087

11. Greenspan FS, Gardner DG (2001) Basic and clinical endocrinology, 6th edn. McGraw Hill, New York

12. Brincat MP (2000) Hormone replacement therapy and the skin. Maturitas 35:107–117

13. Morley JE (2001) Androgens and aging. Maturitas 38:61–73

14. Callens A, Vaillant L, Lecomte P, Berson M, Gall Y, Lorette G (1996) Does hormonal skin aging exist? A study of the influence of different hormone therapy regimens on the skin of postmenopausal women using non-invasive measurement techniques. Dermatology 193:289–294

15. Rudman D, Feller AG, Nagraj HS, Gergans GA, Lalitha PY, Goldberg AF, Schlenker RA, Cohn L, Rudman IW, Mattson DE (1990) Effects of human growth hormone in men over 60 years old. N Engl J Med 323:1–6

16. Raine-Fenning NJ, Brincat MP, Muscat-Baron Y (2003) Skin aging and menopause: implications for treatment. Am J Clin Dermatol 4:371–378

17. Sator PG, Schmidt JB, Rabe T, Zouboulis CC (2004) Skin aging and sex hormones in women – clinical perspectives for intervention by hormone replacement therapy. Exp Dermatol 13 [Suppl 4]:36–40

18. Dieudonne MN, Pecquery R, Leneveu MC, Giudicelli Y (2000) Opposite effects of androgens and estrogens on adipogenesis in rat preadipocytes: evidence for sex and site-related specificities and possible involvement of insulin-like growth factor 1 receptor and peroxisome proliferator-activated receptor γ2. Endocrinology 141:649–656

19. Tavakkol A, Varani J, Elder JT, Zouboulis CC (1999) Maintenance of human skin in organ culture: role for insulin-like growth factor-1 receptor and epidermal growth factor receptor. Arch Dermatol Res 291:643–651

20. Ma H, Sprecher HW, Kolattukudy PE (1998) Estrogen-induced production of a peroxisome proliferator-activated receptor (PPAR) ligand in a PPARγ-expressing tissue. J Biol Chem 273:30131–30138

21. Rosenfield RL, Kentsis A, Deplewski D, Ciletti N (1999) Rat preputial sebocyte differentiation involves peroxisome proliferator-activated receptors. J Invest Dermatol 112:226–232

22. Miller KK (2001) Androgen deficiency in women. J Clin Endocrinol Metab 86:2395–2401

23. Makrantonaki E, Oeff MK, Fimmel S, Zouboulis CC (2003) Molecular mechanisms of hormone-induced skin aging. Exp Dermatol 12:333

24. Chen W, Yang C-C, Sheu H-M, Seltmann H, Zouboulis CC (2003) Expression of peroxisome proliferator-activated receptor and CCAAT/enhancer binding protein transcription factors in cultured human sebocytes. J Invest Dermatol 121:441–447

25. Oettel M (2002) Is there a role for estrogens in the maintenance of men's health? Aging Male 5:248–257

26. Katz MS (2000) Geriatric grand rounds: Eve's rib, or a revisionist view of osteoporosis in men. J Gerontol A Biol Sci Med Sci 55:M560–569

27. Brossner C, Petritsch K, Fink K, Auprich M, Madersbacher S, Adlercreutz H, Rehak P, Petritsch P (2004) Phytoestrogen tissue levels in benign prostatic hyperplasia and prostate cancer and their association with prostatic diseases. Urology 64:707–711

28. Ponholzer A, Plas E, Schatzl G, Jungwirth A, Madersbacher S (2002) Austrian Society of Urology. Association of DHEA-S and estradiol serum levels to symptoms of aging men. Aging Male 5:233–238

29. Juul A, Skakkebaek NE (2002) Androgens and the ageing male. Hum Reprod Update 8:423–433

30. Von Eckardstein S, Nieschlag E (2000) Therapie mit Sexualhormonen beim alternden Mann. Dt Ärztebl 97:A3175–3182

31. Kao JS, Garg A, Mao-Qiang M, Crumrine D, Ghadially R, Feingold KR, Elias PM (2001) Testosterone perturbs epidermal permeability barrier homeostasis. J Invest Dermatol 116:443–451

32. Gilliver SC, Wu F, Ashcroft GS (2003) Regulatory roles of androgens in cutaneous wound healing. Thromb Haemost 90:978–985

33. Gambera A, Scagliola P, Falsetti L, Sartori E, Bianchi U (2004) Androgens, insulin-like growth factor-I (IGF-I), and carrier proteins (SHBG, IGFBP-3) in postmenopause. Menopause 11:159–166

34. Ghizzani A, Razzi S, Fava A, Sartini A, Picucci K, Petraglia F (2003) Management of sexual dysfunctions in women. J Endocrinol Invest 26 [Suppl 3]:137–138

35. Buvat J (2003) Androgen therapy with dehydroepiandrosterone. World J Urol 21:346–355

36. Hennessey JV, Chromiak JA, DellaVentura S, Reinert SE, Puhl J, Kiel DP, Rosen CJ, Vandenburgh H,

MacLean DB (2001) Growth hormone administration and exercise effects on muscle fiber type and diameter in moderately frail older people. J Am Geriatr Soc 49:852–858

37. Kann PH (2003) Growth hormone in anti-aging medicine: a critical review. Aging Male 6:257–263
38. Harman SM, Blackman MR (2004) Use of growth hormone for prevention or treatment of effects of aging. J Gerontol A Biol Sci Med Sci 59:652–658
39. Savine R, Sonksen P (2000) Growth hormone – hormone replacement for the somatopause? Horm Res 53 [Suppl 3]:37–41
40. Mauras N, Rini A, Welch S, Sager B, Murphy SP (2003) Synergistic effects of testosterone and growth hormone on protein metabolism and body composition in prepubertal boys. Metabolism 52:964–969
41. Prakasam G, Yeh JK, Chen MM, Castro-Magana M, Liang CT, Aloia JF (1999) Effects of growth hormone and testosterone on cortical bone formation and bone density in aged orchiectomized rats. Bone 24:491–497
42. Lissett CA, Shalet SM (2003) The insulin-like growth factor-I generation test: peripheral responsiveness to growth hormone is not decreased with ageing. Clin Endocrinol (Oxf) 58:238–245
43. Weber MM (2002) Effects of growth hormone on skeletal muscle. Horm Res 58 [Suppl 3]:43–48
44. Akamatsu H, Zouboulis CC, Orfanos CE (1992) Control of human sebocyte proliferation in vitro by testosterone and 5-alpha-dihydrotestosterone is dependent on the localization of the sebaceous glands. J Invest Dermatol 99:509–511
45. Deplewski D, Rosenfield RL (1999) Growth hormone and insulin-like growth factors have different effects on sebaceous cell growth and differentiation. Endocrinology 140:4089–4094
46. Zouboulis CC, Makrantonaki E (2003) Molekulare Mechanismen der hormonellen Hautalterung. J Dtsch Dermatol Gesellschaft 1 [Suppl 1]:S94
47. Ruiz-Torres A, Soares de Melo Kirzner M (2002) Ageing and longevity are related to growth hormone/insulin-like growth factor-1 secretion. Gerontology 48:401–407
48. Lee KS, Oh KY, Kim BC (2000) Effects of dehydroepiandrosterone on collagen and collagenase gene expression by skin fibroblasts in culture. J Dermatol Sci 23:103–110
49. Baulieu E-E, Thomas G, Legrain S, Lahlou N, Roger M, Debuire B, Faucounau V, Girard L, Hervy M-P, Latour F, Leaud M-C, Mokrane A, Pitti-Ferrandi H, Trivalle C, de Lacharrière O, Nouveau S, Rakoto-Arison B, Souberbielle J-C, Raison J, Le Bouc Y, Raynaud A, Girerd X, Forette F (2000) Dehydroepiandrosterone (DHEA), DHEA sulfate, and aging: contribution of the DHEAge Study to a sociobiomedical issue. Proc Natl Acad Sci U S A 97:4279–4284
50. Percheron G, Hogrel JY, Denot-Ledunois S, Fayet G, Forette F, Baulieu EE, Fardeau M, Marini JF (2003) Double-blind placebo-controlled trial. Effect of 1-year oral administration of dehydroepiandrosterone to 60- to 80-year-old individuals on muscle function and cross-sectional area: a double-blind placebo-controlled trial. Arch Intern Med 163:720–727
51. Genazzani AR, Inglese S, Lombardi I, Pieri M, Bernardi F, Genazzani AD, Rovati L, Luisi M (2004) Long-term low-dose dehydroepiandrosterone replacement therapy in aging males with partial androgen deficiency. Aging Male 7:133–143
52. Allolio B, Arlt W (2002) DHEA treatment: myth or reality? Trends Endocrinol Metab 13:288–294
53. Huber J, Gruber C (2001) Immunological and dermatological impact of progesterone. Gynecol Endocrinol 15 [Suppl 6]:18–21
54. Wieser F, Hosmann J, Tschugguel W, Czerwenka K, Sedivy R, Huber JC (2001) Progesterone increases the number of Langerhans cells in human vaginal epithelium. Fertil Steril 75:1234–1235
55. Mingjia L, Short R (2002) How oestrogen or progesterone might change a woman's susceptibility to HIV-1 infection. Aust N Z J Obstet Gynaecol 42:472–475
56. Fischer TW, Elsner P (2001) The antioxidative potential of melatonin in the skin. Curr Probl Dermatol 29:165–174
57. Abdel-Wahhab MA, Abdel-Galil MM, El-Lithey M (2005) Melatonin counteracts oxidative stress in rats fed an ochratoxin A contaminated diet. J Pineal Res 38:130–135
58. Fischer TW, Greif C, Fluhr JW, Wigger-Alberti W, Elsner P (2004) Percutaneous penetration of topically applied melatonin in a cream and an alcoholic solution. Skin Pharmacol Physiol 17:190–194
59. Fischer T, Bangha E, Elsner P, Kistler GS (1999) Suppression of UV-induced erythema by topical treatment with melatonin. Influence of the application time point. Biol Signals Recept 8:132–135
60. Wespes E, Schulman CC (2002) Male andropause: myth, reality, and treatment. Int J Impot Res 14 [Suppl 1]:S93–98
61. Traustadottir T, Bosch PR, Matt KS (2005) The HPA axis response to stress in women: effects of aging and fitness. Psychoneuroendocrinology 30:392–402
62. Kaplan MM, Pan CY, Gordon PR, Lee JK, Gilchrest BA (1988) Human epidermal keratinocytes in culture convert thyroxine to 3,5,3′-triiodothyronine by type II iodothyronine deiodination: a novel endocrine function of the skin. J Clin Endocrinol Metab 66:815–822
63. Morrow LB (1978) How thyroid disease presents in the elderly. Geriatrics 33:42–45

64. Mirza FS, Prestwood KM (2004) Bone health and aging: implications for menopause. Endocrinol Metab Clin North Am 33:741–759

65. Vieth R, Ladak Y, Walfish PG (2003) Age-related changes in the 25-hydroxyvitamin D versus parathyroid hormone relationship suggest a different reason why older adults require more vitamin D. J Clin Endocrinol Metab 88:185–191

66. Dawson-Hughes B (2004) Racial/ethnic considerations in making recommendations for vitamin D for adult and elderly men and women. Am J Clin Nutr 80 [Suppl 6]:1763S–1766S

67. Zouboulis CC (2004) The human skin as a hormone target and an endocrine gland. Hormones 3:9–26

68. Cao T, Tsai SY, O'Malley BW, Wang XJ, Roop DR (2002) The epidermis as a bioreactor: topically regulated cutaneous delivery into the circulation. Hum Gene Ther 13:1075–1080

69. Lemmey AB, Glassford J, Flick-Smith HC, Holly JM, Pell JM (1997) Differential regulation of tissue insulin-like growth factor-binding protein (IGFBP)-3, IGF-I and IGF type 1 receptor mRNA levels, and serum IGF-I and IGFBP concentrations by growth hormone and IGF-I. J Endocrinol 154:319–328

70. Zouboulis CC, Degitz K (2004) Androgen action on human skin – from basic research to clinical significance. Exp Dermatol 13 [Suppl 4]:5–10

6

Chapter 7

Permeability and Antioxidant Barriers in Aged Epidermis

7

Jens Thiele, Chantal O. Barland, Ruby Ghadially, Peter M. Elias

Contents

7.1 Introduction: Protective Functions of the Skin

Both the cutaneous permeability and the antioxidant barriers, which are critical for life in a hostile, terrestrial environment, largely reside in the outer layers of the epidermis, the stratum corneum (SC). Though traditionally viewed as inert, the SC is both metabolically active and interactive with the underlying epidermal cell layers [17]. The lipid-depleted corneocytes of the SC are embedded in a lipid-enriched extracellular matrix which is organized into characteristic lamellar membranes, and it is this lipid/protein complex that together determines both barrier integrity and antioxidant defense [16, 65] (Table 7.1). The lipid-enriched component consists predominantly of ceramides (about 50% by weight), as well as cholesterol and free fatty acids (FFA) [72]. Importantly, these three species comprise about 10% of the dry weight of "young" SC, where they are deployed in an approximate molar ratio of 1:1:1 [50].

Because it is located at the interface with the environment, the SC is directly exposed to an onslaught of prooxidative stressors, including air pollutants, ultraviolet solar radiation (UVR), chemical oxidants, and microorganisms [8, 78, 79, 87]. To counteract oxidative injury, human skin is equipped with a network of enzymatic and non-enzymatic antioxidant systems [87] (Fig. 7.1). An imbalance between reactive oxygen species (ROS) and antioxidant protection mechanisms leads to oxidation of macromolecules, including DNA, lipids, and proteins. Oxidative modifications of biomolecules may result in loss of structural and/or functional integrity of key components of the epidermal barrier.

Table 7.1. How lipids mediate key stratum corneum defensive functions

Characteristic	Permeability	Antioxidant	Antimicrobial	Hydration
Extracellular localization (only intercellular lipids impact function)	+	+	+	+
Amount of lipid (wt%)	+	?	?	+
Elongated, tortuous pathway (increases diffusion path length)	+	–	+	–
Organization into lamellar membranes	+	–	–	+
Hydrophobic composition (absence of polar lipids and presence of very long-chain saturated fatty acids)	+	+	–	–
Correct molar ratio (approximately 1:1:1 of three key lipids: ceramides, cholesterol, and FFA)	+	–	–	–
Unique molecular structures (e.g., acylceramides)	+	–	–	–
Added sebum constituents, including glycerol and vitamin E	–	+	+	+

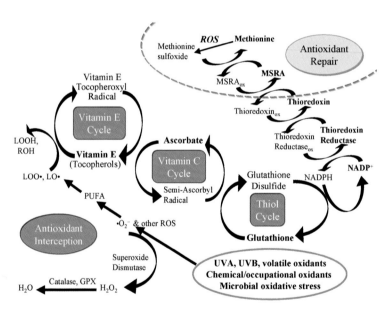

Fig. 7.1. The antioxidant network of human skin: free radical interception and antioxidant repair (*GPX* glutathione peroxidases, *LOO·* lipid peroxyl radical, *LOOH* lipid hydroperoxides, *MSRA* methionine sulfoxide reductase, *ROS* reactive oxygen species)

7.2 Molecular Basis for Epidermal Aging

Despite the many well-characterized morphological, biophysical, and biochemical features of the SC, epidermal redox properties are only beginning to be studied systematically. On a cellular level, apoptosis and decreased proliferative capacity, two key features of aging, affect epidermal structure and function, while on a molecular level, telomere shortening and ROS, respectively, account for the characteristic structural and functional changes of chronologically aged skin [24, 91]. While excess ROS from mitochondrial oxidative metabolism mediate many of the changes seen in chronological aging, UVR gen-

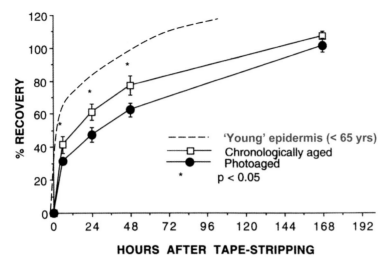

Fig. 7.2. Barrier recovery is delayed in chronologically and photoaged skin (modified from Ref. 66)

erates superoxide anions, hydrogen peroxide, hydroxyl radicals, and singlet oxygen at various levels of human skin, including extra- and intracellular components of the epidermis and dermis [73, 81]. UVR-induced ROS generation is dependent on the wavelength, which determines the depth of penetration into the skin, as well as on the quantity and quality of photosensitizers present. As opposed to volatile oxidants, which may oxidize skin surface components, UVR is considered to be pathophysiologically far more relevant for oxidative skin injury and altered barrier function [88].

Telomeres, which shorten with chronological aging, are repetitive DNA sequences that cap the ends of chromosomes. Since terminal base pairs are lost with each round of mitosis, progressive chromosome shortening occurs with each somatic cell division [91]. Yet UV-induced DNA damage also leads to telomere destabilization and exposure of a single-stranded 3′ overhang of telomeric DNA. Destabilization of the loop structure by either UV-induced DNA damage, and/or progressive depletion with chronological aging, exposes the telomeric repeat sequence. Then, an as-yet-unidentified sensing mechanism interacts with the overhang to initiate a signal cascade that can lead to cell cycle arrest, premature senescence, or apoptosis [91].

7.3 Structural Basis for the Aged Barrier

The phenotypic appearance of aged skin results from the accumulation of functional alterations from chronological aging, and additive morphological/physiological changes from chronic photodamage [24, 91]. For example, significant hyperplasia develops only in sun-exposed skin [15]. The permeability barrier abnormality (Fig. 7.2), however, can be attributed to reduced delivery of secreted lipids to the SC [27]. The decrement in secretion, in turn, results in a reduction in the number of extracellular lamellar membranes in the SC, as visualized by electron microscopy. Superimposition of photoaging further aggravates the barrier abnormality [66] (Fig. 7.2). As a result, the extracellular matrix of aged SC is more porous to both water and xenobiotes than is young epidermis [39, 88].

7.4 Biochemical Basis for Permeability Barrier Abnormality in Aged Skin

Although the epidermal lipid synthetic apparatus is highly active and relatively autonomous from circulating influences, it can be regulated by changes in barrier function [21]. Barrier perturbations stimulate the synthesis of all three key lipids, which fuels the formation and accelerated secretion of new lamellar bodies and rapid replenishment of lipid in the SC interstices for the barrier [31, 53]. Acute barrier disruption also coordinately increases the mRNA levels of several enzymes required for cholesterol, fatty acid, and ceramide synthesis [37]. The coordinate increase in the mRNA levels for the enzymes of cholesterol and FA synthesis is due, in turn, to regulation of these genes by sterol regulatory element binding proteins (SREBPs) [38]. In contrast, while ceramide synthesis is not regulated directly by SREBPs, serine palmitoyl transferase (SPT) activity and ceramide levels could still be regulated indirectly by SREBPs through the availability of bulk fatty acids for formation of either the sphingoid base and/or N-acyl fatty acid moieties. SPT is a functional heterodimer in which the SPT1 and SPT2 subunits combine to catalyze the initial rate-limiting step in the synthesis of ceramides; i.e., condensation of L-serine and palmitoyl CoA [35]. While SPT mRNA protein expression and activity also increase in response to inflammatory stimuli, including UV light [20], increased SPT activity here, however, reflects ceramide signaling of apoptosis [89]. Thus, photoinduced upregulation of SPT activity, leading to ceramide-mediated apoptosis, could negatively impact barrier homeostasis.

Although several laboratories have described reduced content and synthesis of ceramides in aged SC [10], levels of cholesterol and FFA, are also reduced [27]. This global reduction in aged SC lipids (about one-third less than in young SC) explains the paucity of membrane structures that underlie the barrier deficit in aged SC (see above). The decline in ceramide, cholesterol, and fatty acid content/synthesis is attributable, in turn, to reduced activities of the key (rate-limiting) enzymes for each of these lipids including SPT, HMGCoA reductase, and acetyl CoA carboxylase (ACC) [28]. But the most profound abnormality is in cholesterol synthesis; hence, the ability of topical cholesterol-dominant lipid mixtures to normalize function in aged epidermis (see section 7.9).

7.5 Abnormal Signaling of Metabolic Responses in Aged Epidermis

7.5.1 Calcium Gradient and Calcium Signaling

Whereas low levels of calcium are normally present in the lower epidermis, extracellular and intracellular calcium levels peak in the outer stratum granulosum in young epidermis [25, 51]. The epidermal calcium gradient serves at least two key functions: (1) the induction of terminal differentiation, and (2) regulation of exocytosis of lamellar bodies [18]. In aged epidermis, however, the calcium gradient is largely lost (Fig. 7.3), with calcium distributed more evenly throughout the epidermis [11], due to a decreased number or activity of ion pumps, ion channels, and/or ionotropic receptors in aged skin. While the role of Ca^{2+} in lamellar body secretion is known, it has been further postulated that an altered calcium gradient in aged epidermis could account for the barrier abnormality [11].

7.5.2 Cytokines and Growth Factors

While the lipid biochemical basis for the barrier abnormality in aged epidermis is known, abnormal autocrine/paracrine signaling could account for diminished lipid synthesis [30]. Cytokines, such as tumor necrosis factor α (TNFα), interleukin-1 α (IL-1α), IL-6 and interferon α (IFNα), increase de novo synthesis of ceramides, fatty acids and cholesterol in the liver [22, 33, 52]. Moreover, keratinocytes also synthesize diverse growth factors, including several members of the epidermal growth factor (EGF) family, platelet-derived growth factor, nerve growth factor (NGF), and vascular endothelial growth factor (VEGF), which function as autocrine regulators of growth and differentiation [42, 48]. In addition, systemic administration of NGF increases serum triglyceride and FFA levels [58]. The downstream effects of the EGF family members, including EGF itself, TGFα, and amphiregulin (AR), include both stimulation of DNA synthesis and phosphorylation of several cytoplasmic proteins, including the ribosomal

CONSEQUENCES OF SIGNALLING ABNORMALITIES IN INTRINSICALLY AGED EPIDERMIS

Fig. 7.3. Consequences of abnormal cytokine signaling in aged skin

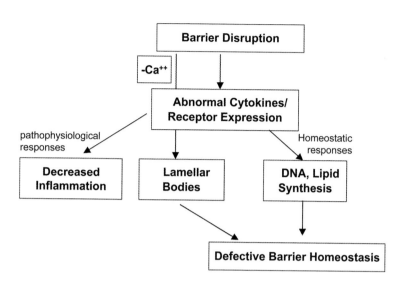

protein, S6 [62]. The receptor's tyrosine kinase activity regulates expression of c-fos and c-myc, epidermal growth, and DNA synthesis [57, 62]. AR is a potent autocrine stimulator of keratinocyte proliferation and lipid synthesis, processes that are signaled through cell surface glycosaminoglycans; i.e., heparin inhibits the mitogenic response of cultured cells to exogenous AR [7, 62].

Although the EGF-related proteins share certain biological activities, AR appears to be unique among EGF family members for its apparent link to barrier homeostasis; i.e. it is the only EGF family member that is upregulated by barrier requirements [47], and pertinently, AR expression declines with age [63, 92]. Since IL-1α stimulates epidermal lipid synthesis in cultured human keratinocytes [1], reduced IL-1α also could contribute to the decline in permeability function in aged skin. Indeed, IL-1α levels decrease in various tissues of the aged, including a decrease in cytokine protein in aged human keratinocytes [70], and lower cytokine mRNA levels in aged mouse and human epidermis [29, 92]. Moreover, IL-1ra expression decreases in photoaging [29]. Furthermore, aged transgenic mice with a knockout of the functional

IL-1α receptor display a barrier abnormality that is not shown by age-matched wild-type littermates [92]. Finally, topical treatment of aged skin with the immunomodulator, imiquimod, enhances (normalizes) barrier recovery rates in aged animals in parallel with increased IL-α production [1]. Together, these results suggest that dysregulation of cytokine production and/or downstream reactivity are general features of aged epidermis that could affect function in aging skin.

7.6 Oxidative Stress

7.6.1 General Mechanisms

The formation of ROS during exposure to UVR is initiated following absorption by endogenous or exogenous chromophores in the skin; e.g., trans-urocanic acid, melanins, flavins, porphyrins, quinones, tryptophan moieties, and glycation end-products [8, 78]. Following absorption, the activated chromophore may react in two ways. In type I photoreactions, the excited chromophore directly reacts with a substrate molecule via electron or hydrogen atom trans-

7

fer, giving rise to free radicals. In the presence of molecular oxygen (minor type II reactions), the initial type I reaction is followed by formation of the superoxide anion radical ($\cdot O_2^-$). Subsequently, $\cdot O_2^-$ generates hydrogen peroxide (H_2O_2) either by spontaneous dismutation or catalytically via superoxide dismutase (Fig. 7.1). Alternatively, in the presence of metal ions such as Fe(II) or Cu(II), H_2O_2 can be converted to the highly reactive hydroxyl radical ($\cdot OH$). In major type II reactions, reactive singlet oxygen ($^1O_2$) is formed from UVR-excited chromophores in the presence of triplet oxygen $^3O_2$ (molecular oxygen in its ground state). ROS species can then react with an array of macromolecules, including lipids, proteins, carbohydrates and DNA. Unsaturated lipids that react with ROS form lipid peroxyl (LOO$\cdot$) and alkoxyl radicals (LO$\cdot$), which tend to initiate chain-propagating, autocatalytic reactions ("lipid peroxidation"). These lipid peroxidation end-products, such as malondialdehyde (MDA) or 4-hydroxynonanol (4-HNE), induce cellular stress responses and, at higher concentrations, they are cytotoxic.

7.6.2 Oxidation of Proteins

ROS, whether generated as by-products of cellular metabolism or whether they are of environmental origin, can modify amino acid structure sufficiently to result in loss of function. Oxidation can also fragment polypeptide chains directly, and crosslink peptides and proteins [75]. Protein carbonyls, which serve as a marker of ROS-mediated protein oxidation, are formed either by oxidative cleavage of proteins, or by direct oxidation of lysine, arginine, proline, or threonine residues [3]. Finally, ROS can also cause specific "fingerprint" modifications of amino acids that result in changes in structural and enzymatic proteins. Methionine residues, in proteins are particularly susceptible to oxidation to methionine sulfoxide (MetSO), and repair of this damage by methionine sulfoxide reductases (MSRA) is critical for cellular survival in the presence of ROS [75] (Fig. 7.1).

Analogous to the high degree of saturation of fatty acids in SC, the extent of disulfide crosslinking of proteins in human SC is many-fold

higher than in lower epidermal layers [6]. Keratins in the SC typically contain dramatically more carbonyl groups than the same keratins in cultured keratinocytes, indicating that levels of keratin oxidation are considerably higher in the SC than in lower epidermal layers [6, 84]. A steep gradient of carbonyl groups in SC keratins is present, with low levels in the lower SC, and high levels in the outer layers of human SC. Importantly, this protein oxidation gradient correlates inversely with the gradients of vitamin E [83], and reduced free thiols [6] found in human SC. Because proteases are involved in epidermal desquamation [13] and selectively hydrolyze oxidized proteins [75], Thiele and coworkers have proposed that epidermal protein oxidation and its reversal by MSRA may be involved in the regulation of desquamation [84].

7.7 Impact of Oxidative Stress on the Stratum Corneum

The SC is the skin layer most susceptible to volatile environmental pollutants, such as ozone [80–82, 88, 90]. However, since ozone is so reactive, it reacts primarily with lipids and proteins in the outer SC. While UVR, including ultraviolet B (UVB; 280–320 nm) and ultraviolet A (UVA; 320–400 nm) is capable of penetrating into deeper skin layers initiating photooxidative stress, skin surface lipids and the SC are exposed to the highest UVB and UVA doses. Nevertheless, little is known about the impact of UVR-induced photooxidative stress in the skin barrier. While immediate oxidative damage is found in SC components upon UVR exposure, functional barrier disturbances are generally detected not earlier than 24-48 hours after exposure [36, 88]. Those changes presumably reflect oxidant damage to lipids and/or proteins [43, 77, 84]. However, they may also involve inflammatory cascades initiated in nucleated epidermal layers. The prime mechanism of both UVA- and UVB-induced oxidative tissue damage is thought to be the peroxidation of lipids, which is inhibited by antioxidants, such as vitamin E [78, 87] (Table 7.1). Yet total SC antioxidant activity, and particularly vitamin E levels, decline dramatically following exposure to solar-simulated

UVR, even at low physiological UVA and UVB doses [83]. The steep decline in SC vitamin E may be due to the paucity of other antioxidants in the SC, which normally recycle each other [76] (Fig. 7.1). Indeed, a lack of hydrophilic antioxidants, such as ascorbic acid, uric acid, and reduced glutathione, would be expected in the SC because of the hydrophobic environment in this tissue.

Skin surface lipids, in particular squalene, which coat the surface of sebaceous gland-enriched skin sites, such as the face, are first-line targets of UVR. Irradiation of squalene with both UVB and UVA has been reported to induce squalene hydroperoxide formation [44, 60, 67]. However, at suberythemogenic doses, UVA generates squalene monohydroperoxides (SqmOOH) at rates at least one order of magnitude higher than UVB, and thus appears to be of higher relevance [14]. The inverse correlation found between squalene depletion and SqmOOH formation further confirms that natural sebaceous gland-derived squalene is the substrate for SqmOOH formation following UVA exposure. Yet, because squalene is also a good quencher of singlet oxygen [44], it may also serve as an early line of antioxidant defense.

7.8 Effects of Chronological Aging and Photoaging on Antioxidant Defense

Several key metabolic enzymes are oxidatively inactivated by mixed function oxidases during chronological aging. These inactivated enzymes accumulate, but only after the age of 60 years [59]. Yet individuals with premature aging syndromes, such as progeria or Werner's syndrome, display significantly higher levels of oxidized proteins than age-matched controls [59]. Another study has demonstrated that aged fibroblasts accumulate increased oxidized proteins upon oxidative stress and are less efficient in removing oxidized proteins [54]. Thus, loss of functional enzyme activity of enzymes during aging may be due to oxidative modification by mixed function oxidation systems.

There is in vivo evidence that photoaged skin exhibits high levels of different markers of oxidative stress, including the accumulation of lipid peroxides, glycation end-products, and oxidized proteins [41, 68, 69, 77]. Following chronic UV exposure, significantly increased levels of oxidatively modified proteins are found both in the papillary dermis and in the outermost layers of the SC [68, 84]. The fact that nucleated epidermal layers are less affected by protein oxidation probably reflects both UV filtration by the SC and the far greater antioxidant capacity of the epidermis, including a stronger expression of the protein oxidation repair enzyme MSRA (Fig. 7.1). Remarkably, catalase protein and activity levels are naturally very high in human SC and become significantly depleted upon UVR exposure in vivo [40, 68].

7.9 Physiological Antioxidants in the Epidermal Barrier

7.9.1 Introduction

Living organisms have developed complex, integrated extracellular and intracellular defenses against oxidative stress. Plant and animal epithelial surfaces and respiratory tract surfaces contain antioxidants that provide defense against environmental stress caused by ambient ROS, thus ameliorating their injurious effects on underlying cellular constituents [8]. An overwhelming of antioxidant systems at these biosurfaces may result in the formation of specific oxidation products, some of which have been identified as biomarkers of environmental oxidative stress. Moreover, secondary or tertiary reaction products may also induce injury to underlying cells or be involved in physiological tissue and site-specific responses such as epidermal differentiation and desquamation.

7.9.2 Non-enzymatic Antioxidants in the Stratum Corneum

Non-enzymatic antioxidant concentrations as well as enzymatic antioxidant activities are generally significantly higher in the epidermis than in the dermis of murine and human skin

[86]. This probably reflects the fact that the epidermis is more directly exposed to various exogenous sources of oxidative stress and thus requires a higher antioxidant defense capacity to maintain a physiological redox balance. On a molar basis, hydrophilic non-enzymatic antioxidants including L-ascorbic acid, GSH and uric acid appear to be the predominant antioxidants in human skin. Their overall dermal and epidermal concentrations are more than 10- to 100-fold greater than those found for vitamin E or ubiquinol. In contrast to uric acid, GSH and ubiquinol, vitamins C and E cannot be synthesized by the human body and must be taken up by the diet. Consequently, the skin's antioxidant defense may be at least partially influenced by nutritive factors. However, to date little is known about the physiological regulation of antioxidant vitamins in human skin [86].

α-Tocopherol is known to efficiently scavenge lipid peroxyl and alkoxyl radicals by intercepting lipid chain propagation. The initial oxidation product of tocopherol is the meta-stable tocopheroxyl radical, which can either react with another lipid radical leading to α-tocopherol consumption, or abstracts a hydrogen atom from polyunsaturated lipids to give α-tocopherol and lipid radical. In the second case, occurring preferentially at low lipid radical concentrations, the lipid radical may later react with oxygen to form a lipid peroxyl radical. As demonstrated in vitro in lipid and cellular systems, when ascorbic acid or ubiquinol ("Coenzyme Q10") are present, the tocopheroxyl radical is rapidly reduced and the active form of vitamin E, α-tocopherol is regenerated. Hence, the lack of such "co-antioxidants" from the antioxidant network may result in limited antioxidant protection of lipid bilayers or other lipophilic domains. Remarkably, a vitamin E gradient is present in the SC of untreated, healthy human skin, with the lowest tocopherol concentrations at the surface and highest in the deepest SC layers. Analogous to the vitamin E gradient in SC, concentrations of ascorbate are lowest in the outer SC, increasing almost tenfold in the lower SC [86, 87]. However, even the lower SC displays ascorbate levels that are at least one order of magnitude lower than epidermal ascorbate levels. Urate levels are

higher and urate is distributed more evenly in human SC than are ascorbate and tocopherols with only a small increase towards deeper layers of SC [87]. Finally, similar gradients, with the highest levels in lower SC layers, have been found for ascorbate, glutathione and uric acid in hairless mouse SC [90].

In summary, while hydrophilic antioxidants are overall less abundant in the hydrophobic SC environment than in nucleated epidermal layers, vitamin E appears to be the predominant antioxidant in human SC. Depletion of SC vitamin E is one of the earliest and most sensitive oxidative stress markers in human skin exposed to UVR and non-UVR stresses in vivo [87].

7.9.3 Sebum Delivery of Vitamin E

The first studies on sebum antioxidants were based on the rather unexpected observation that the upper SC layers of environmentally exposed human facial skin contain severalfold higher levels of α-tocopherol than corresponding layers of less-exposed upper arm SC [85]. Since a film of skin surface lipids that largely originate from sebaceous gland secretion covers the facial SC, we hypothesized that sebaceous gland secretion provides a conduit for vitamin E delivery to the skin surface. In fact, α-tocopherol levels are 20-fold higher on facial than on forearm skin, and levels decrease progressively with SC depth. It is not yet known whether vitamin E continues to be delivered to the skin surface in equivalent quantities in aged skin, with or without carriage in sebum. Remarkably, levels of α-tocopherol in human sebum are more than threefold higher than levels in plasma [46], dermis, epidermis [74], or SC [83]. As shown for other sebum lipids, sebaceous α-tocopherol penetrates into subjacent SC layers, where it accounts for the increased levels of α-tocopherol detected in the SC of the sebaceous gland-enriched skin sites [87]. Thus, sebaceous gland secretion provides a physiological pathway for delivery of α-tocopherol. Similar pathways have been described for drug delivery of certain lipid-soluble oral antifungal compounds [19].

7.9.4 Enzymatic Antioxidants in the Stratum Corneum

Few data are available on the presence and activity of enzymatic antioxidants, such as catalase, superoxide dismutases (present in human skin as the manganese form and the copper–zinc form), and glutathione peroxidases in the SC. The major role of catalase as an antioxidant is its ability to detoxify H_2O_2 by decomposing two H_2O_2 molecules to two molecules of water and one of oxygen (Fig. 7.1). In most tissues, highest catalase activities are found in the peroxisomes, where it constitutes about 50% of the peroxisomal protein. Remarkably high levels of catalase activity appear to persist in human SC [34]. We have recently found high protein levels of catalase, and, to a lesser extent, of superoxide dismutases in human SC [68]. Remarkably, catalase levels are significantly decreased in acutely or chronically UV-exposed skin, as well as in intrinsically aged skin. These results were confirmed by Hellemans et al. on the activity level of the enzyme [40]. The same authors also showed that SC catalase levels and activity are strongly affected by UVA, but remain unchanged upon UVB exposure. Remarkably, lesional skin of patients with polymorphic light eruption (PLE) display significantly decreased SC catalase levels [34]. Some authors consider UVA-induced photo-oxidation to be involved in the pathogenesis of PLE. Consequently, topical pretreatment of skin with antioxidants has recently been suggested as a new treatment approach for moderate-to-severe PLE [23].

Besides "interceptive" antioxidant systems that intercept free radical accumulation and thus prevent oxidative damage, numerous human tissues are equipped with "repair enzymes" that are able to reverse and thus control protein oxidative damage. An antioxidant repair enzyme that has received considerable attention lately is the peptide MSRA, which was initially identified in *Escherichia coli* and has also been identified in a number of different human tissues [56]. We have recently described the presence of MSRA in human skin, where it is most strongly expressed in basal and suprabasal epidermis [69]. MSRA reverses the inactivation of many proteins due to oxidation of critical methionine (Met) residues by reducing methionine sulfoxide (MetO) to Met (Fig. 7.1). Unlike other antioxidant enzymes, its action is independent of metals or cofactors, but it does require reducing equivalents generated from a thioredoxin regenerating system, and such a system (i.e., thioredoxin and thioredoxin reductase) is present in human epidermis [71]. Whether or not such repair enzymes of oxidative protein damage are involved in the regulation of epidermal pathways controlling desquamation or counteract the aging process remains to be elucidated.

7.10 Barrier Function and Repair in Aged Epidermis

7.10.1 Permeability Barrier Restitution

Although aged epidermis displays normal basal barrier function, barrier recovery kinetics are delayed after acute insults [28], with a further delay in photoaged skin [66] (Fig. 7.2). Based upon the lipid biochemical abnormalities in aged epidermis, we assessed which type of SC lipid mixtures could correct the functional abnormality. In young murine or human skin, any incomplete mixture of one or two of the three major lipid species (cholesterol, ceramides, and FFA) worsens barrier function (Table 7.2) [50]. In contrast, equimolar mixtures of the three key lipids allow normal rates of barrier recovery in young skin (Table 7.2). Further adjustment of the three-component mixtures to a molar ratio of 3:1:1 actually accelerates barrier recovery significantly [49], and in young skin each of the three key species can predominate. In contrast, in aged epidermis, with its global decline in lipid synthesis and profound abnormality in cholesterol synthesis, the requirements for barrier repair are quite different. Topical cholesterol alone, which delays barrier recovery in young skin, accelerates barrier recovery in aged murine and human skin (Table 7.2) [28]. Moreover, while equimolar mixtures of the three key lipids also accelerate barrier recovery in aged epidermis, optimized ratios only accelerate recovery if cholesterol is the predominant species (Fig. 7.4)

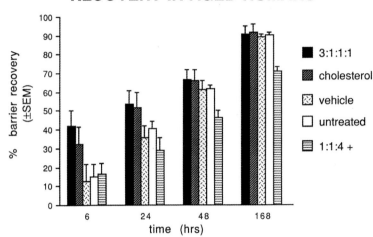

Fig. 7.4. Cholesterol must be the dominant lipid for optimal barrier recovery in aged humans (modified from Ref. 93).

Table 7.2. Effects of physiological lipid mixtures on barrier recovery in young and aged human skin (physiological lipids: FFA, cholesterol, ceramides)

Physiological lipids	Young	Aged
Single lipid	Delays	Accelerates (cholesterol only)
Triple lipids (equimolar)	No change	Accelerates
Triple lipids (optimized)		
Fatty acid-dominant	Accelerates	Delays
Ceramide-dominant	Accelerates	Not studied
Cholesterol-dominant	Accelerates	Accelerates

[93][1]. These results underscore the selective (profound) abnormality in cholesterol synthesis that characterizes the aged epidermal permeability barrier.

7.10.2 Antioxidant Barrier Restitution

The skin is equipped with two photoprotective barriers: a melanin barrier in the lower epidermis, and a protein/urocanic acid barrier in the SC, which absorbs significant amounts of UVB. With respect to UVB-induced erythema, SC thickness appears to be more important for photoprotection than either the extent of pigmentation or the total thickness of viable epidermis [31]. Only few studies on the mechanisms of action of the photoprotective properties of the SC are currently available. Recent studies on the redox properties of the SC point to a relevance of photooxidative changes in the barrier and the protective action of SC antioxidants. Topically applied antioxidants provide protection against UVB-induced oxidative damage in SC lipids [60]. Even some systemically applied antioxidants accumulate in the SC and play an important role against UV-induced

---

[1] The cholesterol-dominant mixture, shown to repair the barrier of aged human skin, is available commercially (Crème de l'Extrème, Osmotics Corporation, Denver, Colo.).

photodamage in skin. Studies on the photoprotective mechanisms of the antioxidant butylated hydroxytoluene (BHT) suggest that changes in the physicochemical properties of SC keratins occur, leading to increases in UV-absorption of underlying epidermal layers. It has been proposed that these changes are exerted via the anti-radical action of BHT that retards oxidation and prevents crosslinking of the keratin chains, resulting in a diminution of UVB radiation reaching potential epidermal target sites [4]. Similar mechanisms of action are currently discussed for other lipophilic antioxidants such as tocopherols and carotenoids [86].

7.10.2.1 Lipid-Soluble Antioxidants

As was shown for synthetic antioxidant compounds such as BHT, physiological lipophilic antioxidants such as tocopherols and carotenoids have been shown to possess photoprotective properties [86]. Vitamin E may also stabilize SC lamellar membranes [60] since the degree of disorder and the amount of lipids decrease in the outer cell layers of human SC [5]. For better stability, vitamin E is commonly used as a biologically non-active esterified form. Vitamin E acetate, succinate and linoleate, are also capable of reducing certain endpoints of UVR-induced skin damage [2, 45]. However, they appear to be less effective than unesterified vitamin E. Vitamin E esters act as prodrugs since they are hydrolyzed to the active, free vitamin E (α-tocopherol) upon penetration into skin. There is conflicting evidence as to the extent to which this conversion actually takes place in the SC. In human SC, the bioconversion rate of vitamin E esters into vitamin E is lower than in nucleated epidermal layers. Therefore, α-tocopherol should provide more efficient antioxidant protection of skin surface lipids and skin barrier constituents than vitamin E esters.

7.10.2.2 Water-Soluble Antioxidants

Photoprotective effects of topically applied vitamin C have been shown in several studies [9, 12]. Unlike vitamin E, however vitamin C does not absorb in the UV range, and therefore it does not have sunscreen properties. Yet topically applied vitamin C is photoprotective when formulated at high concentrations in an appropriate vehicle [9], and at a low pH [64]. The rather modest photoprotective effect of topically applied vitamin C in human skin may also be explained, in part, by its instability and ease of oxidation in aqueous vehicles [26]. Yet vitamin C can be protected from degradation by selecting appropriate vehicles such as emulsions [26]. Furthermore, lipophilic and more stable vitamin C derivatives, such as its palmityl, succinyl or phosphoryl esters, have been reported to be promising compounds providing greater photoprotection than vitamin C. However, as described for vitamin E esters, most of these compounds must be hydrolyzed to vitamin C to be effective as antioxidants, and thus are unlikely to be active in protecting the SC. Furthermore, although lipophilic vitamin C derivatives are more stable than vitamin C, in vitro studies on UVB-irradiated keratinocytes suggest that ascorbyl palmitate exerts a net prooxidizing effect [55].

According to the antioxidant network theory (Fig. 7.1), combinations of synergistic coantioxidants such as vitamins E and C may help to avoid prooxidative free radical damage that may occur upon UVR exposure after topical application of high doses of a single antioxidant compound. Furthermore, as outlined in this chapter, there is growing experimental evidence that cutaneous photoprotection should involve antioxidant strategies to prevent damage to SC lipids and proteins and skin surface lipids.

References

1. Barland CO, Zettersten E, Brown BS, Ye J, Elias PM, Ghadially R (2004) Imiquimod-induced interleukin-1 alpha stimulation improves barrier homeostasis in aged murine epidermis. J Invest Dermatol 122:330–336
2. Baschong W, Artmann C, Hueglin D, Roeding J (2001) Direct evidence for bioconversion of vitamin E acetate into vitamin E: an ex vivo study in viable human skin. J Cosmet Sci 52:155–161
3. Berlett BS, Stadtman ER (1997) Protein oxidation in aging, disease, and oxidative stress. J Biol Chem 272:20313–20316

7

4. Black HS, Mathews-Roth MM (1991) Protective role of butylated hydroxytoluene and certain carotenoids in photocarcinogenesis. Photochem Photobiol 53:707–716

5. Bommannan D, Potts RO, Guy RH (1990) Examination of stratum corneum barrier function in vivo by infrared spectroscopy. J Invest Dermatol 95:403–408

6. Broekaert D, Cooreman K, Coucke P, Nsabumukunzi S, Reyniers P, Kluyskens P, Gillis E (1982) A quantitative histochemical study of sulphydryl and disulphide content during normal epidermal keratinization. Histochem J 14:573–584

7. Cook PW, Mattox PA, Keeble WW, Shipley GD (1992) Inhibition of autonomous human keratinocyte proliferation and amphiregulin mitogenic activity by sulfated polysaccharides. In Vitro Cell Dev Biol 28A:218–222

8. Cross CE, van der Vliet A, Louie S, Thiele JJ, Halliwell B (1998) Oxidative stress and antioxidants at biosurfaces: plants, skin, and respiratory tract surfaces. Environ Health Perspect 106 [Suppl 5]:1241–1251

9. Darr D, Combs S, Dunston S, Manning T, Pinnell S (1992) Topical vitamin C protects porcine skin from ultraviolet radiation-induced damage. Br J Dermatol 127:247–253

10. Denda M, Koyama J, Hori J, Horii I, Takahashi M, Hara M, Tagami H (1993) Age- and sex-dependent change in stratum corneum sphingolipids. Arch Dermatol Res 285:415–417

11. Denda M, Tomitaka A, Akamatsu H, Matsunaga K (2003) Altered distribution of calcium in facial epidermis of aged adults. J Invest Dermatol 121:1557–1558

12. Dreher F, Gabard B, Schwindt DA, Maibach HI (1998) Topical melatonin in combination with vitamins E and C protects skin from ultraviolet-induced erythema: a human study in vivo. Br J Dermatol 139:332–339

13. Egelrud T, Lundstrom A, Sondell B (1996) Stratum corneum cell cohesion and desquamation in maintenance of the skin barrier. In: Marzulli F, Maibach H (eds) Dermatotoxicology. Taylor & Francis, Washington, pp 19–27

14. Ekanayake Mudiyanselage S, Hamburger M, Elsner P, Thiele JJ (2003) Ultraviolet A induces generation of squalene monohydroperoxide isomers in human sebum and skin surface lipids in vitro and in vivo. J Invest Dermatol 120:915–922

15. El-Domyati M, Attia S, Saleh F, Brown D, Birk DE, Gasparro F, Ahmad H, Uitto J (2002) Intrinsic aging vs. photoaging: a comparative histopathological, immunohistochemical, and ultrastructural study of skin. Exp Dermatol 11:398–405

16. Elias PM (1983) Epidermal lipids, barrier function, and desquamation. J Invest Dermatol [Suppl] 80:44s–49s

17. Elias PM, Wood LC, Feingold KR (1999) Epidermal pathogenesis of inflammatory dermatoses. Am J Contact Dermat 10:119–126

18. Elias PM, Ahn SK, Denda M, Brown BE, Crumrine D, Kimutai LK, Komuves L, Lee SH, Feingold KR (2002) Modulations in epidermal calcium regulate the expression of differentiation-specific markers. J Invest Dermatol 119:1128–1136

19. Faergemann J, Godleski J, Laufen H, Liss RH (1995) Intracutaneous transport of orally administered fluconazole to the stratum corneum. Acta Derm Venereol 75:361–363

20. Farrell AM, Uchida Y, Nagiec MM, Harris IR, Dickson RC, Elias PM, Holleran WM (1998) UVB irradiation up-regulates serine palmitoyltransferase in cultured human keratinocytes. J Lipid Res 39:2031–2038

21. Feingold KR (1991) The regulation and role of epidermal lipid synthesis. Adv Lipid Res 24:57–82

22. Feingold KR, Soued M, Serio MK, Moser AH, Dinarello CA, Grunfeld C (1989) Multiple cytokines stimulate hepatic lipid synthesis in vivo. Endocrinology 125:267–274

23. Fesq H, Ring J, Abeck D (2003) Management of polymorphous light eruption: clinical course, pathogenesis, diagnosis and intervention. Am J Clin Dermatol 4(6):399–406

24. Fisher GJ, Kang S, Varani J, Bata-Csorgo Z, Wan Y, Datta S, Voorhees JJ (2002) Mechanisms of photoaging and chronological skin aging. Arch Dermatol 138:1462–1470

25. Forslind B, Werner-Linde Y, Lindberg M, Pallon J (1999) Elemental analysis mirrors epidermal differentiation. Acta Derm Venereol 79:12–17

26. Gallarate M, Carlotti ME, Trotta M, Bovo S (1999) On the stability of ascorbic acid in emulsified systems for topical and cosmetic use. Int J Pharm 188:233–241

27. Ghadially R, Brown BE, Sequeira-Martin SM, Feingold KR, Elias PM (1995) The aged epidermal permeability barrier. Structural, functional, and lipid biochemical abnormalities in humans and a senescent murine model. J Clin Invest 95:2281–2290

28. Ghadially R, Brown BE, Hanley K, Reed JT, Feingold KR, Elias PM (1996) Decreased epidermal lipid synthesis accounts for altered barrier function in aged mice. J Invest Dermatol 106:1064–1069

29. Gilchrest BA, Garmyn M, Yaar M (1994) Aging and photoaging affect gene expression in cultured human keratinocytes. Arch Dermatol 130:82–86

30. Gilhar A, Aizen E, Pillar T, Eidelman S (1992) Response of aged versus young skin to intradermal administration of interferon gamma. J Am Acad Dermatol 27:710–716

31. Gniadecka M, Wulf HC, Mortensen NN, Poulsen T (1996) Photoprotection in vitiligo and normal skin. A quantitative assessment of the role of stratum corneum, viable epidermis and pigmentation. Acta Derm Venereol 76:429–432

32. Grubauer G, Elias PM, Feingold KR (1989) Transepidermal water loss: the signal for recovery of barrier structure and function. J Lipid Res 30:323–333

33. Grunfeld C, Soued M, Adi S, Moser AH, Dinarello CA, Feingold KR (1990) Evidence for two classes of cytokines that stimulate hepatic lipogenesis: relationships among tumor necrosis factor, interleukin-1 and interferon-alpha. Endocrinology 127:46–54

34. Guarrera M, Ferrari P, Rebora A (1998) Catalase in the stratum corneum of patients with polymorphic light eruption. Acta Derm Venereol 78:335–336

35. Hanada K, Hara T, Nishijima M (2000) Purification of the serine palmitoyltransferase complex responsible for sphingoid base synthesis by using affinity peptide chromatography techniques. J Biol Chem 275:8409–8415

36. Haratake A, Uchida Y, Mimura K, Elias PM, Holleran WM (1997) Intrinsically aged epidermis displays diminished UVB-induced alterations in barrier function associated with decreased proliferation. J Invest Dermatol 108:319–323

37. Harris IR, Farrell AM, Grunfeld C, Holleran WM, Elias PM, Feingold KR (1997) Permeability barrier disruption coordinately regulates mRNA levels for key enzymes of cholesterol fatty acid, and ceramide synthesis in the epidermis. J Invest Dermatol 109:783–787

38. Harris IR, Farrell AM, Holleran WM, Jackson S, Grunfeld C, Elias PM, Feingold KR (1998) Parallel regulation of sterol regulatory element binding protein-2 and the enzymes of cholesterol and fatty acid synthesis but not ceramide synthesis in cultured human keratinocytes and murine epidermis. J Lipid Res 39:412–422

39. Harvell JD, Maibach HI (1994) Percutaneous absorption and inflammation in aged skin: a review. J Am Acad Dermatol 31:1015–1021

40. Hellemans L, Corstjens H, Neven A, Declercq L, Maes D (2003) Antioxidant enzyme activity in human stratum corneum shows seasonal variation with an age-dependent recovery. J Invest Dermatol 120:434–439

41. Jeanmaire C, Danoux L, Pauly G (2001) Glycation during human dermal intrinsic and actinic ageing: an in vivo and in vitro model study. Br J Dermatol 145:10–18

42. Karvinen S, Pasonen-Seppanen S, Hyttinen JM, Pienimaki JP, Torronen K, Jokela TA, Tammi MI, Tammi R (2003) Keratinocyte growth factor stimulates migration and hyaluronan synthesis in the epidermis by activation of keratinocyte hyaluronan synthases 2 and 3. J Biol Chem 278:49495–49504

43. Kitazawa M, Podda M, Thiele J, Traber MG, Iwasaki K, Sakamoto K, Packer L (1997) Interactions between vitamin E homologues and ascorbate free radicals in murine skin homogenates irradiated with ultraviolet light. Photochem Photobiol 65:355–365

44. Kohno Y, Egawa Y, Itoh S, Nagaoka S, Takahashi M, Mukai K (1995) Kinetic study of quenching reaction of singlet oxygen and scavenging reaction of free radical by squalene in n-butanol. Biochim Biophys Acta 1256:52–56

45. Kramer-Stickland K, Liebler DC (1998) Effect of UVB on hydrolysis of alpha-tocopherol acetate to alpha-tocopherol in mouse skin. J Invest Dermatol 111:302–307

46. Lang JK, Gohil K, Packer L (1986) Simultaneous determination of tocopherols, ubiquinols, and ubiquinones in blood, plasma, tissue homogenates, and subcellular fractions. Anal Biochem 157:106–116

47. Liou A, Elias PM, Grunfeld C, Feingold KR, Wood LC (1997) Amphiregulin and nerve growth factor expression are regulated by barrier status in murine epidermis. J Invest Dermatol 108:73–77

48. Maas-Szabowski N, Starker A, Fusenig NE (2003) Epidermal tissue regeneration and stromal interaction in HaCaT cells is initiated by TGF-alpha. J Cell Sci 116:2937–2948

49. Man MM, Feingold KR, Thornfeldt CR, Elias PM (1996) Optimization of physiological lipid mixtures for barrier repair. J Invest Dermatol 106:1096–1101

50. Mao-Qiang M, Feingold KR, Elias PM (1993) Exogenous lipids influence permeability barrier recovery in acetone-treated murine skin. Arch Dermatol 129:728–738

51. Mauro T, Bench G, Sidderas-Haddad E, Feingold K, Elias P, Cullander C (1998) Acute barrier perturbation abolishes the Ca2+ and K+ gradients in murine epidermis: quantitative measurement using PIXE. J Invest Dermatol 111:1198–1201

52. Memon RA, Holleran WM, Moser AH, Seki T, Uchida Y, Fuller J, Shigenaga JK, Grunfeld C, Feingold KR (1998) Endotoxin and cytokines increase hepatic sphingolipid biosynthesis and produce lipoproteins enriched in ceramides and sphingomyelin. Arterioscler Thromb Vasc Biol 18:1257–1265

53. Menon GK, Feingold KR, Elias PM (1992) Lamellar body secretory response to barrier disruption. J Invest Dermatol 98:279–289

54. Merker K, Sitte N, Grune T (2000) Hydrogen peroxide-mediated protein oxidation in young and old human MRC-5 fibroblasts. Arch Biochem Biophys 375:50–54

55. Meves A, Stock SN, Beyerle A, Pittelkow MR, Peus D (2002) Vitamin C derivative ascorbyl palmitate promotes ultraviolet-B-induced lipid peroxidation and cytotoxicity in keratinocytes. J Invest Dermatol 119:1103–1108

56. Moskovitz J, Bar-Noy S, Williams WM, Requena J, Berlett BS, Stadtman ER (2001) Methionine sulfoxide reductase (MsrA) is a regulator of antioxidant defense and lifespan in mammals. Proc Natl Acad Sci U S A 98:12920–12925

57. Nanney LB, Sundberg JP, King LE (1996) Increased epidermal growth factor receptor in fsn/fsn mice. J Invest Dermatol 106:1169–1174

58. Nonogaki K, Moser AH, Shigenaga J, Feingold KR, Grunfeld C (1996) Beta-nerve growth factor as a mediator of the acute phase response in vivo. Biochem Biophys Res Commun 219:956–961

59. Oliver CN, Ahn BW, Moerman EJ, Goldstein S, Stadtman ER (1987) Age-related changes in oxidized proteins. J Biol Chem 262:5488–5491

60. Pelle E, Muizzuddin N, Mammone T, Marenus K, Maes D (1999) Protection against endogenous and UVB-induced oxidative damage in stratum corneum lipids by an antioxidant-containing cosmetic formulation. Photodermatol Photoimmunol Photomed 15:115–119

61. Picardo M, Zompetta C, De Luca C, Cirone M, Faggioni A, Nazzaro-Porro M, Passi S, Prota G (1991) Role of skin surface lipids in UV-induced epidermal cell changes. Arch Dermatol Res 283:191–197

62. Piepkorn M, Lo C, Plowman G (1994) Amphiregulin-dependent proliferation of cultured human keratinocytes: autocrine growth, the effects of exogenous recombinant cytokine, and apparent requirement for heparin-like glycosaminoglycans. J Cell Physiol 159:114–120

63. Piepkorn M, Underwood RA, Henneman C, Smith LT (1995) Expression of amphiregulin is regulated in cultured human keratinocytes and in developing fetal skin. J Invest Dermatol 105:802–809

64. Pinnell SR, Yang H, Omar M, Monteiro-Riviere N, DeBuys HV, Walker LC, Wang Y, Levine M (2001) Topical L-ascorbic acid: percutaneous absorption studies. Dermatol Surg 27:137–142

65. Rawlings AV, Scott IR, Harding CR, Bowser PA (1994) Stratum corneum moisturization at the molecular level. J Invest Dermatol 103:731–740

66. Reed JT, Elias PM, Ghadially R (1997) Integrity and permeability barrier function of photoaged human epidermis. Arch Dermatol 133:395–396

67. Saint-Leger D, Bague A, Lefebvre E, Cohen E, Chivot M (1986) A possible role for squalene in the pathogenesis of acne. II. In vivo study of squalene oxides in skin surface and intra-comedonal lipids of acne patients. Br J Dermatol 114:543–552

68. Sander CS, Chang H, Salzmann S, Muller CS, Ekanayake-Mudiyanselage S, Elsner P, Thiele JJ (2002) Photoaging is associated with protein oxidation in human skin in vivo. J Invest Dermatol 118:618–625

69. Sander CS, Hansel A, Heinemann SH, Elsner P, Thiele JJ (2002) In vivo evidence for a link between photoaging and oxidative stress in human skin. J Invest Dermatol 119:331A

70. Sauder DN, Ponnappan U, Cinader B (1989) Effect of age on cutaneous interleukin 1 expression. Immunol Lett 20:111–114

71. Schallreuter KU, Wood JM (2001) Thioredoxin reductase – its role in epidermal redox status. J Photochem Photobiol B 64:179–184

72. Schurer NY, Elias PM (1991) The biochemistry and function of stratum corneum lipids. Adv Lipid Res 24:27–56

73. Shindo Y, Witt E, Packer L (1993) Antioxidant defense mechanisms in murine epidermis and dermis and their responses to ultraviolet light. J Invest Dermatol 100:260–265

74. Shindo Y, Witt E, Han D, Packer L (1994) Dose-response effects of acute ultraviolet irradiation on antioxidants and molecular markers of oxidation in murine epidermis and dermis. J Invest Dermatol 102:470–475

75. Stadtman ER (2001) Protein oxidation in aging and age-related diseases. Ann N Y Acad Sci 928:22–38

76. Stoyanovsky DA, Osipov AN, Quinn PJ, Kagan VE (1995) Ubiquinone-dependent recycling of vitamin E radicals by superoxide. Arch Biochem Biophys 323:343–351

77. Tanaka N, Tajima S, Ishibashi A, Uchida K, Shigematsu T (2001) Immunohistochemical detection of lipid peroxidation products, protein-bound acrolein and 4-hydroxynonenal protein adducts, in actinic elastosis of photodamaged skin. Arch Dermatol Res 293:363–367

78. Thiele JJ (2001) Oxidative targets in the stratum corneum. A new basis for antioxidative strategies. Skin Pharmacol Appl Skin Physiol 14 [Suppl 1]:87–91

79. Thiele JJ, Podda M, Packer L (1997) Tropospheric ozone: an emerging environmental stress to skin. Biol Chem 378:1299–1305

80. Thiele JJ, Traber MG, Podda M, Tsang K, Cross CE, Packer L (1997) Ozone depletes tocopherols and tocotrienols topically applied to murine skin. FEBS Lett 401:167–170

81. Thiele JJ, Traber MG, Polefka TG, Cross CE, Packer L (1997) Ozone-exposure depletes vitamin E and induces lipid peroxidation in murine stratum corneum. J Invest Dermatol 108:753–757

82. Thiele JJ, Traber MG, Tsang K, Cross CE, Packer L (1997) In vivo exposure to ozone depletes vitamins C and E and induces lipid peroxidation in epidermal layers of murine skin. Free Radic Biol Med 23:385–391

83. Thiele JJ, Traber MG, Packer L (1998) Depletion of human stratum corneum vitamin E: an early and sensitive in vivo marker of UV induced photo-oxidation. J Invest Dermatol 110:756–761

84. Thiele JJ, Hsieh SN, Briviba K, Sies H (1999) Protein oxidation in human stratum corneum: susceptibility of keratins to oxidation in vitro and presence of a keratin oxidation gradient in vivo. J Invest Dermatol 113:335–339

85. Thiele JJ, Weber SU, Packer L (1999) Sebaceous gland secretion is a major physiologic route of vitamin E delivery to skin. J Invest Dermatol 113:1006–1010

86. Thiele J, Dreher F, Packer L (2000) Antioxidant defense systems in skin. In: Elsner P, Maibach H (eds) Drugs vs. cosmetics: cosmeceuticals? Marcel Dekker, New York, pp 145–188

87. Thiele JJ, Schroeter C, Hsieh SN, Podda M, Packer L (2001) The antioxidant network of the stratum corneum. Curr Probl Dermatol 29:26–42

88. Thiele JJ, Dreher F, Maibach HI, Packer L (2003) Impact of ultraviolet radiation and ozone on the transepidermal water loss as a function of skin temperature in hairless mice. Skin Pharmacol Appl Skin Physiol 16:283–290

89. Uchida Y, Nardo AD, Collins V, Elias PM, Holleran WM (2003) De novo ceramide synthesis participates in the ultraviolet B irradiation-induced apoptosis in undifferentiated cultured human keratinocytes. J Invest Dermatol 120:662–669

90. Weber SU, Thiele JJ, Cross CE, Packer L (1999) Vitamin C, uric acid, and glutathione gradients in murine stratum corneum and their susceptibility to ozone exposure. J Invest Dermatol 113:1128–1132

91. Yaar M, Eller MS, Gilchrest BA (2002) Fifty years of skin aging. J Investig Dermatol Symp Proc 7:51–58

92. Ye J, Calhoun C, Feingold K, Elias P, Ghadially R (1999) Age-related changes in the IL-1 gene family and their receptors before and after barrier abrogation. J Invest Dermatol 112:543

93. Zettersten EM Ghadially R, Feingold KR, Crumrine D, Elias PM (1997) Optimal ratios of topical stratum corneum lipids improve barrier recovery in chronologically aged skin. J Am Acad Dermatol 37:403–408

Clarifying the Vitamin D Controversy: The Health Benefits of Supplementation by Diet Versus Sunshine

8

Deon Wolpowitz, Barbara A. Gilchrest

Contents

8.1 The Controversy

The suspicion that ultraviolet (UV) radiation is a human carcinogen has existed in the medical literature since at least the 1890s [4]. In spite of this, two discoveries in the 1920s swayed public opinion that unprotected sunshine exposure was beneficial to health and therefore desirable: (1) vitamin D as the active compound in cod-liver oil that prevented childhood rickets (see below) [114], and (2) the role of UV radiation in vitamin D synthesis [4]. This favorable public perception of UV radiation was further reinforced by exaggerated claims of health benefits by some members of the medical community and by those who financially benefited from this favorable attitude – the makers of home UV lamps [4]. At the same time, from the late 1920s through the 1950s, others in the medical community intensified their efforts to increase public awareness of the carcinogenic nature of UV radiation. In the 1960s, UV radiation-induced DNA mutations were characterized, and in the 1990s, specific genes and signaling pathways targeted by these mutations were identified [4].

There is now overwhelming evidence that UV radiation is carcinogenic. The legacy from the first half of the 20th century is a near epidemic of melanoma and nonmelanoma skin cancers. In 2000, an estimated 1.3 million skin cancers were diagnosed in the United States [97]. Approximately 1 in 70 people in the United States are expected to develop malignant melanoma during their lifetime [96], and the incidence of all types of skin cancer continues to increase. Also, the adverse psychosocial and physiological effects of photoaging fuel the demand for effective interventions among the progressively increasing absolute number and proportion of the population who are elderly [116]. Neverthe-

Table 8.1. Nomenclature of vitamin D metabolites and enzymes

Common name	Clinical name/formula	Abbreviation
7-Dehydrocholesterol	ProvitaminD$_3$	7-DHC
Cholecalciferol	Vitamin D$_3$	Vit-D$_3$
Ergocalciferol	Vitamin D$_2$	Vit-D$_2$
Calcidiol	25-Hydroxyvitamin D	25-OH vit D
Calcitriol	1,25-Dihydroxyvitamin D	1,25-(OH)$_2$ vit D
–	24,25-Dihydroxyvitamin D	24,25-(OH)$_2$ vit D
–	1α,24,25-Trihydroxyvitamin D	1α,24,25-(OH)$_3$ vit D
CYP27A1	Vitamin D-25-hydroxylase	–
CYP27B1	25-OH(D)-1α-hydroxylase	–
CYP24	24-Hydroxylase	–

Table 8.2. Formulas to convert different units of measurement

Daily vitamin D intake	Micrograms (µg) ×40 = international units (IU)
Serum 25-OH vitamin D	Nanomoles per liter (nmol/l) ×0.4 = nanograms per milliliter (ng/ml)
Serum PTH	Picograms per milliliter (pg/ml) ×0.11 = picomoles per liter (pM/l)

less, society still maintains that tanned skin is cosmetically attractive, and a multi-billion dollar a year tanning industry exists to capitalize on this perception [28, 117]. Clearly, further efforts are needed to reduce unprotected sun exposure, increase sun protection, and increase sun avoidance to reverse this trend in the general population.

Unfortunately, reducing sunlight (UVA and UVB) exposure curtails UVB-dependent vitamin D production (see Figs. 8.1 and 8.5). However, as detailed below, some within the medical community currently advocate increasing baseline vitamin D stores in the general public. The question therefore arises whether the general public should be encouraged to obtain higher vitamin D stores, and if so whether by augmenting cutaneous production, when this entails exposure to a known carcinogen, or via diet and/or oral supplements.

This chapter reviews the nomenclature, metabolism, established and putative functions of vitamin D; the definition, significance, and known clinical consequences of vitamin D "deficiency" versus "insufficiency"; the populations at risk of vitamin D deficiency; and the risks versus benefits of obtaining vitamin D from UV exposure versus diet or oral supplementation.

8.2 Nomenclature and Sources

Vitamin D obtained its name in the early part of the 20th century following the discovery of the antirachitic effect of cod-liver oil. The suspected vitamin in cod-liver oil was designated "D" as vitamins A, B, and C had already been identified [114]. The nomenclature and units of measurements associated with vitamin D metabolites and enzymes are confusing, and a summary is provided in Tables 8.1 (nomenclature) and 8.2 (units). The physiological production and metabolism of vitamin D metabolites is summarized briefly in Fig. 8.1, and has been extensively reviewed elsewhere (for references see Ref. 47). Salient points are summarized here. First, the term "vitamin D" specifically refers to two biologically inert precursors: vitamin D$_2$ (ergocalciferol) and vitamin D$_3$ (cholecalciferol) [46, 112]. Vitamins D$_2$ and D$_3$ can both enter the circulation through the gastrointestinal tract

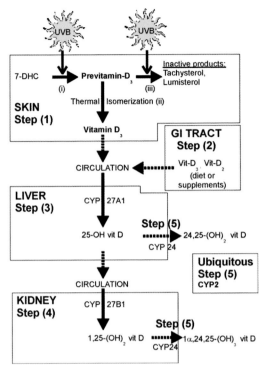

Fig. 8.1. Sources, sites, and processing of vitamin D metabolites. In the skin (step 1), 7-DHC is converted by UVB into previtamin D_3, that further UVB exposure converts them to the biologically inactive products tachysterol and lumisterol. In this way, tachysterol and lumisterol formation acts as a biological safety valve to prevent UVB-induced vitamin D toxicity. Thermal isomerization converts previtamin D_3 into vitamin D_3 which then enters the circulation. Vitamin D_3/vitamin D_2 also enter the circulation by absorption through the gastrointestinal tract (step 2). Vitamin D_3/vitamin D_2, bound to vitamin D-binding protein, are transported to the liver (step 3) where CYP27A1 converts them to biologically inactive 25-OH vitamin D. 25-OH vitamin D is carried to the kidney (step 4) where it is converted to 1,25-$(OH)_2$ vitamin D, the biologically active metabolite of vitamin D metabolism. Finally, both 25-OH vitamin D and 1,25-$(OH)_2$ vitamin D are metabolized to biologically inactive products by a ubiquitous enzyme, CYP24

(Fig. 8.1, step 2), but the conventional American diet is low in vitamins D_2/D_3, naturally found in some yeasts and plants, fatty salt-water fish, fish (cod) liver oils, and eggs [47, 85]. However, both vitamins D_2/D_3 are also available in fortified foods, including milk, orange juice, infant formulas, and in pill-form as nonprescription dietary supplements and prescription tablets [110].

Second, cutaneous production of vitamin D_3 is absolutely dependent on UV radiation present in sunlight in the UVB spectrum (290–320 nm) [45, 54]. Continued UVB exposure is not a limitless source of vitamin D as additional UVB further transforms previtamin-D_3 into biologically inactive metabolites, tachysterol and lumisterol (Fig. 8.1, step 1(iii)) [70]. In fact, cutaneous vitamin D_3 production to a single prolonged UVB stimulation is capped at approximately 10–15% of the original 7-DHC concentration [45] and is achieved with suberythemogenic UV exposures [53].

Third, vitamins D_2/D_3 from the diet via intestinal absorption (Fig. 8.1, step 2) or vitamin D_3 from cutaneous UVB-mediated synthesis (Fig. 8.1, step 1) are both biologically inactive and require subsequent hydroxylation reactions in the liver and kidney to form 1,25-dihydroxyvitamin D (1,25-$(OH)_2$ vitamin D; calcitriol), the biologically active metabolite of vitamin D metabolism (Fig. 8.1, steps 3 and 4) [112]. Thus, dietary consumption and cutaneous vitamin D production are completely interchangeable sources of precursors of the active hormone of vitamin D metabolism.

8.3 Established Hormonal Function and the Consequences of Deficiency or Excess

The principal physiological function of the active vitamin D metabolite, 1,25-$(OH)_2$ vitamin D, is to maintain calcium homeostasis. In this essential role, vitamin D metabolites function mechanistically like a hormone. Thus, active vitamin D metabolites are synthesized at sites (liver and kidney) far from the sites of their biological action (gastrointestinal tract/bone) and reach these distant sites through the bloodstream [112]. Figure 8.2 outlines the interactions between serum calcium, parathyroid hormone (PTH), and 1,25-$(OH)_2$ vitamin D in the blood, kidney, bones, and gastrointestinal tract to preserve normal serum calcium levels (for references see Ref. 47).

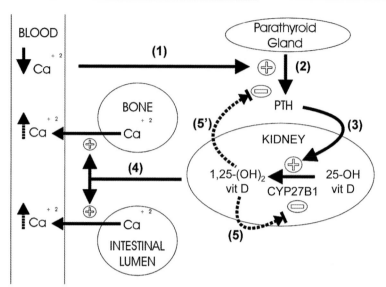

8

Fig. 8.2. Established role of vitamin D metabolites in maintaining serum calcium (Ca^{+2}) homeostasis. *(1)* Cells in the parathyroid gland sense low levels of serum Ca^{+2}. *(2)* In response, parathyroid gland cells secrete parathyroid hormone (PTH). *(3)* In the kidney, PTH (and phosphorus) stimulate CYP27B1 (25-OH D-1-alpha-hydroxylase) to produce more 1,25-$(OH)_2$ vitamin D. *(4)* 1,25-$(OH)_2$ vitamin D then acts on the gastrointestinal tract (mainly small intestine/duodenum) to increase absorption of dietary Ca^{+2} and phosphorus. 1,25-$(OH)_2$ vitamin D also acts on bone to increase mobilization of Ca^{+2} stored in the matrix. Finally, 1,25-$(OH)_2$ vitamin D acts in several negative feedback loops to fine-tune its ability to maintain serum calcium homeostasis. These include inhibiting CYP27B1 in the kidney (as does calcium) *(5)*, inhibiting PTH synthesis in the parathyroid *(5')*, and increasing the activity of CYP24 (the enzyme that inactivates 1,25-$(OH)_2$ vitamin D) in the kidney, cartilage, and intestines [10, 81]

By preserving serum levels of calcium and phosphate, 1,25-$(OH)_2$ vitamin D maintains the calcium×phosphate product in a range that provides osteoblasts with the building blocks required to mineralize the collagen matrix [49, 98]. In this regard, the clinical consequences of hormonal vitamin D deficiency are primarily due to the associated elevation in PTH, called secondary hyperparathyroidism (defined as PTH levels greater than 65 pg/ml), outlined in Fig. 8.3. In effect, PTH compensates for the absence of intestinal derived calcium at the expense of building blocks of the bone matrix – mobilizing calcium stores from bone into the blood and wasting phosphorus in the urine [47, 62].

In turn, insufficient amounts of calcium and/or phosphate deposition in the bone matrix leads to the disorders of inadequate bone matrix (osteoid) mineralization associated with severe vitamin D-deficient states – rickets in children and osteomalacia in adults. Clinically, rickets affects both the growth plate (epiphysis) and the newly formed bone and presents with hypotonia, muscle weakness, prominence of the costochondral junction (rachitic rosary), indentation of the lower ribs (Harrison's groove), increased frequency of fractures, waddling gait from back deformities (kyphosis and lordosis) and bowing deformity of the long bones, skull abnormalities (softened calvarium, parietal flattening, frontal bossing), delayed eruption of permanent dentition, enamel defects, and tetany in severe cases [62]. Osteomalacia in adults can be asymptomatic but also presents as vague, long-standing diffuse bone and muscle pain and skeletal muscle weakness, pelvic deformities, and a waddling gait [62, 92, 93].

Conversely, the clinical consequences of excessive vitamin D are secondary to hypercalcemia and hypercalciuria. Initial symptoms of "hypervitaminosis D" include weakness, lethargy, headaches, nausea, and polyuria. Hy-

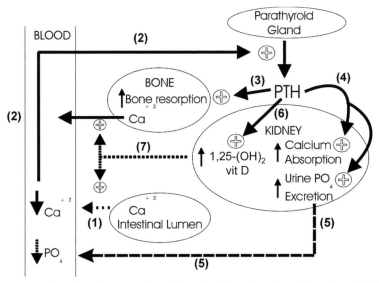

Fig. 8.3. Secondary hyperparathyroidism. Significantly decreased intestinal calcium absorption from vitamin D deficiency *(1)* leads to persistently elevated serum PTH (PTH >65 pg/ml) [72] to defend serum calcium *(2)*. In bone, PTH stimulates osteoclast formation leading to increased bone resorption *(3)*. In the kidney, PTH increases renal tubular absorption of calcium and urinary excretion of phosphorus *(4)*, leading to hypophosphatemia *(5)*. Finally, PTH stimulates CYP27B1 (the enzyme in the kidney that makes 1,25-(OH)$_2$ vitamin D) increasing serum 1,25-(OH)$_2$ vitamin D levels *(6)*, attempting to maximize intestinal calcium absorption and further mobilizing calcium stores from bone *(7)*

percalciuria leads to dehydration and prerenal azotemia. Decreased renal clearance of calcium from prerenal azotemia, in combination with the enhanced gastrointestinal absorption of calcium from elevated 1,25-(OH)$_2$ vitamin D levels, further exacerbates the hypercalcemia. The only route left for calcium excretion becomes ectopic calcification in tissues such as the kidney (nephrocalcinosis, nephrolithiasis), skin, blood vessels (atherosclerosis, supravalvular aortic stenosis), heart, and lungs. In addition, the abnormally high levels of serum calcium lead to mental status changes including confusion, stupor, and coma [98].

8.4 Measuring Metabolites and Defining Deficiency Versus Insufficiency

Vitamin D differs from most other vitamins and essential nutrients in having a large precursor pool in the body that varies over a wide range in healthy individuals and a small tightly regulated range of the active compound. Given the many different vitamin D metabolites, the question arises as to which metabolite most accurately reflects the body's overall vitamin D status. By consensus, 25-OH vitamin D, which has the longest half-life in circulation (approximately 2 or 3 weeks), is the universally accepted indicator of vitamin D status [89, 107]. In addition, 25-OH vitamin D levels vary with intestinal absorption or cutaneous production and so represent the vitamin D intake from both the skin and the diet that enters into the circulation (20–150 nmol/l) [10, 46]. In this regard, 25-OH vitamin D levels show seasonal variation unlike those of 1,25-(OH)$_2$ vitamin D. Finally, 25-OH vitamin D levels correlate with clinical disease states of vitamin D deficiency [2, 94].

Currently, there is no universally accepted measure of "adequate" levels of 25-OH vitamin D [110]. This is because several studies have shown that PTH levels within the normal range (10–65 pg/ml) can be suppressed further by ad-

Table 8.3. Known or purported associations with serum 25-OH vitamin D levels (values in parenthesis indicate associated serum 25-OH vitamin D levels)

Effect on …	Serum 25-OH vitamin D level (nmol/l)			Reference
	Deficient (<20–25)	Insufficient (25–50)	Sufficient (>50)	
Serum PTH level	Above normal range	Within normal range[a]	Maximally suppressed[b]	22, 37, 61, 72, 84, 87, 103, 104, 113
Skeleton	Rickets/osteomalacia	Osteoporosis[c]	Normal skeletal homeostasis	
Prostate cancer risk	Increased (<19)	Baseline (40–60)	Increased (>80)	106
Colorectal adenoma risk	Baseline (<72)	Baseline (<72)	Decreased (>72)	41
Mild essential hypertension (>140/80 mmHg)	Not assessed	Baseline (38–57)	Reduced but still >140/80 mmHg (>100[d])	63
Multiple sclerosis risk	Not assessed	Baseline (40–55)	40% RR reduction (70–75)[e]	80
Fibromyalgia[f]	29% of 150 patients (<20)	93% of 150 patients (<50)	7% of 150 patients (>30)	93

[a] Serum PTH can be further suppressed with higher serum 25-OH vitamin D levels.

[b] Serum PTH levels are no longer correlated to serum 25-OH vitamin D levels above these values.

[c] Defined as increased bone turnover/decreased bone mineral density.

[d] Average 25-OH vitamin D level was 151.2 nmol/l, which causes hypercalciuria – a cause of bone loss and an early sign of hypervitaminosis D.

[e] 40% RR (relative risk) reduction was only found in patients who took multivitamin supplements containing 25-OH vitamin D (and other vitamins) but not in those who achieved similar levels by dietary intake or environmental exposure.

[f] Study assessed 25-OH vitamin D levels in 150 patients with fibromyalgia at only one time point and did not assess if vitamin D supplementation improved symptoms.

ditional supplementation with vitamin D. Terminology is critical to understanding this lack of consensus (Table 8.3). Sufficiency refers to levels of 25-OH vitamin D adequate to prevent abnormalities in calcium homeostasis. Deficiency indicates inadequate levels of 25-OH vitamin D that manifest as clinically evident disease states with respect to skeletal health (i.e., osteomalacia or rickets). In general, levels of 25-OH vitamin D <20–25 nmol/l are associated with deficiency (rickets and histological evidence of osteomalacia). Insufficiency is defined by laboratory values, specifically serum PTH levels within or above the normal range that can be decreased by vitamin D supplementation, in the absence of clinically evident osteomalacia or rickets [92]. More recently, insufficiency has been defined by some as 25-OH vitamin D levels statistically associated with adverse health outcomes in the population such as hypertension, elevated incidence for some internal malignancies, type I diabetes, multiple sclerosis, and fibromyalgia, among other disorders [53] (Table 8.3). In this regard, the case-control and cohort observational studies that have shown a reduction in risk of type I diabetes mellitus with dietary vitamin D supplementation in early childhood do not provide data on corresponding serum 25-OH vitamin D levels [25, 58, 100].

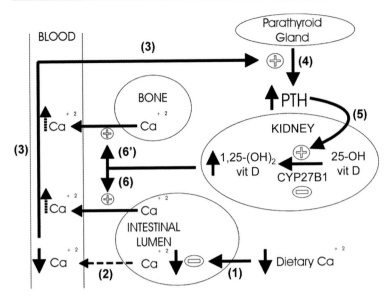

Fig. 8.4. Relationship between a calcium-rich diet, serum 1,25-(OH)₂ vitamin D level, and serum parathyroid hormone (PTH) level. Diets low in calcium decrease the concentration of calcium in the intestinal lumen *(1)*. Low concentrations of calcium in the intestinal lumen decrease the amount of passive absorption of calcium into the bloodstream *(2)*, leading to decreased serum calcium levels *(3)*. Cells in the parathyroid gland sense the decrease in calcium *(3)* and secrete higher baseline levels of PTH to maintain serum calcium homeostasis *(4)*. The baseline higher levels of PTH lead to increased levels of 1,25-(OH)₂ vitamin D *(5)* to augment active absorption of calcium from the gastrointestinal tract *(6)* to compensate for decreased passive absorption *(2)*. Of note, higher levels of PTH also increase mobilization of calcium from bone both directly (see Fig. 8.3) and indirectly through 1,25-(OH)₂ vitamin D *(6')*. The converse relationships exist between diets high in calcium, intestinal calcium absorption, PTH levels, and serum 1,25-(OH)₂ vitamin D levels

Defining the range of insufficiency is not straightforward. The serum 25-OH vitamin D "threshold point" can be defined as the maximum serum 25-OH vitamin D level beyond which there no longer exists an association between further increases in serum 25-OH vitamin D and further decreases in serum PTH. In the elderly, large variations in the serum 25-OH vitamin D threshold point have been found, ranging from as low as 30 nmol/l to greater than 100 nmol/l [22, 37, 61, 72, 87, 104]. This broad range is not likely to be an artifact of different laboratory assays [22]. Variability between laboratories performing the same most frequently used assay is in the range 20–30% [110], which cannot account for a difference of between 30 and 100 nmol/ml.

Instead, Ooms et al. suggest that dietary calcium intake can explain the variation in experimentally determined serum 25-OH vitamin D threshold levels, as an inverse relationship exists between the serum 25-OH vitamin D threshold point and calcium intake [87]. For example, populations with a daily calcium intake greater than 1100 mg versus 700–800 mg a day had serum 25-OH vitamin D threshold points of 30 nmol/l and 110 nmol/l, respectively [22, 87]. Figure 8.4 highlights the inverse relationship between calcium intake and both PTH and 1,25-(OH)₂ vitamin D levels. Importantly, calcium enters the circulation through both vitamin D-dependent active gastrointestinal absorption and also by vitamin D-independent passive absorption [88]. In populations with ample calcium in the diet, vitamin D levels above 30 nmol/ml are not needed because enough calcium is coming into the bloodstream from the gut by passive diffusion, and because higher vitamin D levels are not needed, PTH levels are no longer regulated by vitamin D. Conversely, in populations with lower amounts of calcium in the diet, 25-OH vitamin D levels as high as 100 nmol/ml are needed to maintain serum calcium through active gastrointestinal absorption, and because higher vitamin D levels are needed, PTH levels are higher to generate this compensatory vitamin D.

In summary, the need for the definition of vitamin D insufficiency arose from the assertion that maximal suppression of PTH levels has important health benefits. However, the above data indicate that efforts to suppress PTH maximally should focus on defining adequate levels of dietary calcium intake rather than on arbitrary serum 25-OH vitamin D levels. In fact, in children in equatorial Africa who are vitamin D sufficient, cases of florid rickets have been reported entirely from deficiency of dietary calcium and completely cured by calcium supplementation alone [86]. Thus, because dietary supplementation can simultaneously provide intestinal vitamin D and calcium while sunshine alone provides only vitamin D, dietary supplementation is superior to sunshine alone in suppressing PTH levels – the primary unwanted clinical side effect of vitamin D insufficiency.

8.5 Does Insufficiency Require Supplementation for Skeletal Homeostasis?

Although the clinical benefit of treating insufficiency is not clearly supported by randomized clinical trials of vitamin D supplementation (see below), some authors have chosen to blur the distinction between insufficiency and deficiency. These authors suggest that the low normal range for 25-OH vitamin D should be raised to the highest level that prevents even mild vitamin D insufficiency, as defined by any elevation of PTH [51, 52, 72, 84, 89]. As a consequence of this change in definition, subjects with 25-OH vitamin D levels in the insufficient range are counted as being deficient, and the total number of people in the general population with vitamin D "deficiency" increases significantly. For example, setting the bar for deficiency at 50 nmol/l rather than 25 nmol/l increases the number of vitamin D-"deficient" African–American wom-

en between the ages of 15 and 49 years in the US from 12.2% to 42.4% [84], and the number between the ages of 18 and 29 years in Boston in winter from 6–9% to 30% [103]. Surprisingly, however, in these studies no data were provided on the PTH levels of these subgroups, although in the latter study, the mean PTH for all patients aged 18–50+ years in winter peaked at 44 pg/ml – well below the level that defines secondary hyperparathyroidism [103].

Extending the range of abnormal 25-OH vitamin D levels to include those that cause elevations in PTH but are not associated with osteomalacia or rickets implies that clinical relevance can be ascribed to this laboratory phenomenon alone. As described above, since PTH increases calcium resorption from bone, it has been proposed that vitamin D insufficiency may enhance the process of osteoporosis, especially in the elderly [42, 109]. In support of this concept, prospective randomized trials have demonstrated that vitamin D supplements, even in patients without secondary hyperparathyroidism, decrease bone turnover and increase bone mineral density [20, 21, 23].

Ultimately, however, the most relevant measure of the clinical benefit of preventing secondary hyperparathyroidism, with respect to skeletal metabolism, is the prevention of osteoporosis-related fractures. Prospective observational studies have produced conflicting results regarding the association between vitamin D and/or calcium intake and the primary prevention of osteoporotic fractures in women [27, 79]. Eight prospective, randomized controlled studies of vitamin D metabolites with or without calcium supplementation to prevent osteoporosis-induced fractures have been reported (Table 8.4) [12–15, 23, 29, 43, 69, 78, 105]. These study populations were limited to elderly and mostly female patients (mean age not less than 70 years), restricting the ability to generalize the results to other populations. Of these eight, seven used vitamin D supplements, and one [29] used 1,25-$(OH)_2$ vitamin D. Of the seven studies that used vitamin D supplements, four studies [13, 14, 23, 43, 105] found a statistically significant reduction in fractures with vitamin D supplementation, although caveats pertain, such as a small study

Table 8.4. Efficacy of vitamin D supplementation in preventing osteoporosis-induced fractures and/or increasing serum 25-OH vitamin D levels (*NS* not significant, *NA* not assessed)

Reference	Supplement form	Calcium supplementation per day (mg)	Increase in serum 25-OH vitamin D (nmol/l)	Reduction in hip fractures	Reduction in all non-vertebral fractures
23	D₃ (pill)	500	~29.5[a]	NS	60% RR (P=0.02)
13	D₃ (pill)	1200	65	27%[b] (P=0.004)	26%[b] (P<0.001)
14[e]	D₃ (pill)	1200	NA	23%[b] (P<0.02)	17.2%[b] (P<0.02)
15	D₃ (pill)	1200	30–35	NS	NS
105	D₃ (pill)	None	NA	NS	33% RR (P=0.04)
43	D₂ (IM)	None	18.1 and 31.2[c]	NS	5.4%[d]
78	D₃ (cod-liver oil)	None	22	NS	NS
69	D₃ (pill)	None	31	NS	NS
29	1,25-(OH)₂ vitamin D	None	NS	NA	NS
95	Sunshine[f]	None	18.5	NA	NA
55	UVR (260–320 nm)[g]	None	15.25	NA	NA
17	UVR (270–400 nm)[h]	None	22	None	None

[a] Average of the increases for men and women.

[b] Results of intention-to-treat analysis.

[c] 18.1 nmol/l in outpatients and 31.2 nmol/l in institutionalized patients.

[d] Total number of fractures, including vertebral fractures.

[e] Results of an additional 18 months of follow-up of study population from Ref. 13.

[f] 30 minutes of sunshine; head, neck, forearms, and lower legs exposed; spring (October) in Auckland (latitude 37° South).

[g] National Biologic Light box with SF40 lamps (broad spectrum fluorescent lights).

[h] Philips type PL-S 9W/12 UV-B lamps (spectrum peaking at 310 nm).

size [23], a high proportion of frankly vitamin D-deficient subjects [13, 43], and an improperly randomized or blinded protocol [105]. Moreover, one group found that the intention-to-treat benefit after 3 years of follow-up was less than after 18 months, and although these authors claimed an association, the effect of vitamin D supplementation was not statistically significant in a subsequent confirmation trial [12–15]. Finally, of these five studies, only one study [105] provided solely vitamin D supplementation, whereas the others gave calcium supplementation with vitamin D.

In contrast, two large trials showed no benefit of vitamin D supplementation alone with respect to fracture reduction [69, 78]. Although they employed a lower vitamin D supplementation (400 IU vs 700–800 IU in the positive studies described above), others have reported that increasing intake from 400 to 800 IU marginally affects 25-OH vitamin D levels [68], and a study with 400 IU a day demonstrated a more robust effect on bone mineral density (BMD) at the femoral neck than 700 IU a day [23, 88]. The negative results from these trials [69, 78] have also been attributed to the lower final serum 25-OH concentrations (54 and 64 nmol/l, respectively) versus two positive trials with final serum 25-OH concentrations greater than 100 nmol/l [13, 23]. However, as described in detail above (and see Fig. 8.4), the dietary calcium intake determines the serum 25-OH vitamin D threshold point with respect to PTH suppression, so that final serum 25-OH vitamin D concentrations need to be evaluated in the context of the calcium intake of the patient population [87]. In this regard, the positive study populations at baseline had lower dietary calcium intakes and were the only groups to receive concomitant calcium supplementation. Taken together, the negative results from Meyer et al. [78] and Lips et al. [69] are more likely because fewer of these patients had vitamin D insufficiency, very few had vitamin D deficiency, and overall had higher baseline calcium intakes compared to the study populations of Dawson-Hughes et al. [23] and Chapuy et al. [13].

Finally, within the category of osteoporosis-related fractures, seven of the eight trials summarized above found no statistically significant benefit of vitamin D supplementation in preventing hip fractures (Table 8.4). The one positive study found an intention-to-treat benefit in the first 18 months that was mildly reduced after 36 months and non-significant in a subsequent confirmation trial [12–15]. Consistent with the finding that hip fractures are associated with the highest morbidity and mortality [29], there was no significant difference in mortality rates between treated subjects and controls in two studies that reported this information [43, 105]. Thus, evidence from these randomized, prospective clinical trials suggests that the beneficial effect of vitamin D supplementation (1) requires either concurrent calcium supplementation and/or a diet high in daily calcium, (2) is limited to preventing non-hip osteoporotic fractures, and/or (3) occurs primarily in those patients with severe deficiency (<12 ng/ml = <30 nmol/l) and not in those with insufficiency.

8.6 Other Putative Indications for Treating Insufficiency

8.6.1 Skeletal Muscle Strength?

$1,25\text{-}(OH)_2$ vitamin D, a secosteroid hormone, functions like other steroid hormones. Once inside target cells, $1,25\text{-}(OH)_2$ vitamin D binds to and activates a nuclear receptor, in this case called vitamin D receptor (VDR). This activated complex, which likely contains additional non-VDR receptors, then binds to specific DNA targets (vitamin D-responsive elements) leading to alterations in transcription of effector genes [47, 65]. Thus, VDR-expressing tissues comprise a population of cells theoretically able to respond to $1,25\text{-}(OH)_2$ vitamin D. As would be expected for the classical hormonal function of $1,25\text{-}(OH)_2$ vitamin D in maintaining calcium homeostasis, VDR expression has been documented in intestine, bone, and kidney [101]. In addition, VDR expression has also been documented in brain, skeletal muscle, breast, prostate, colon, activated lymphocytes, macrophages, and skin, creating the possibility that these "non-calcium-homeostasis" tissues might also be regulated by vitamin D [7, 49, 92, 94].

Evidence of the functional role of vitamin D receptors in skeletal muscle and the nervous system comes from closer analysis of the osteoporosis studies described in Table 8.4. Strictly limited to the elderly population, some of these studies found a fracture reduction risk even when bone mineral density was only minimally improved. One explanation for this reduction in osteoporotic fractures is that vitamin D supplementation decreases the number of falls leading to fracture rather than increasing bone strength [111]. Studies have supported the assertion that vitamin D supplementation improves muscle strength and reduces body sway [36, 92]. However, as for osteoporosis-induced fracture reduction, results from clinical trials using vitamin D supplementation to prevent falls have been mixed [8]. This issue turns out to be the same for the fracture studies: in general, studies that showed positive effects in the elderly used vitamin D with calcium supplementation and treated populations that were frankly vitamin D deficient not insufficient [36, 92]. In this regard, a recent meta-analysis found a fall reduction of 22% (corrected OR, 0.78) for five trials that met their specific inclusion criteria and 13% (corrected RR, 0.87) when all relevant studies were pooled. This meta-analysis could not stratify results by baseline 25-OH vitamin D levels but did suggest that the combination of vitamin D and calcium was important for fall reduction [8].

Taken together, the data on fall reduction in the elderly suggest that these modest benefits are both out-weighted by, and possibly unachievable by, cutaneous UVB exposure. In this regard, one cannot conclude definitively that vitamin D supplementation in the absence of concomitant calcium supplementation is effective in preventing falls. Hence, no evidence supports the notion that UVB exposure alone might confer the same benefit as dietary supplementation.

8.6.2 Reduced Cancer Mortality?

Selected epidemiological data suggest an inverse correlation between solar UVB exposure and mortality from several cancers, including colon, breast, and prostate cancer, and between sun exposure and the incidence of colon cancer [30–33, 39, 40] (and for additional references see Refs. 41, 52 and 106). These studies were observational in nature and therefore could not establish that increased solar UVB exposure reduces cancer incidence or mortality. Moreover, much of these data relied on correlating region-specific mortality rates with ambient UV radiation. Such studies may be confounded by geographic variations in cultural or lifestyle behaviors that affect actual sun exposure at the individual level (for example, Ref. 109). In addition, as UV radiation is strongly correlated with latitude, both of these may be confounded by other factors that also vary with geographic location, such as, but not limited to, the earth's magnetic field [76], diet, or pollution.

Nevertheless, cutaneous vitamin D photosynthesis is proposed to account for these epidemiological associations. Observational cross-sectional or case-controlled epidemiological studies have shown an inverse association between vitamin D and these various epithelial derived cancers [31–33, 39, 40] (and for additional references see Refs. 41, 52 and 106). In addition, for human cancer cell lines derived from epithelial and non-epithelial tissues (e.g., prostate, breast, colon, leukemia), 1,25-$(OH)_2$ vitamin D can in vitro be antiproliferative, induce apoptosis, promote cell differentiation, inhibit telomerase expression, and suppress tumor-induced angiogenesis [71] (for specific references see Refs. 41, 52 and 106). Some 1,25-$(OH)_2$ vitamin D analogs have also shown efficacy in vivo in animal models of chemical carcinogenesis [77]. Unfortunately, in general, 1,25-$(OH)_2$ vitamin D levels need to be in the toxic range (i.e., causing hypervitaminosis D) to show these in vitro and in vivo effects [78, 112]. Furthermore, other high-quality epidemiological and observational studies exist that do not support a role for vitamin D in preventing these cancers (for references see Refs. 41, 102 and 106). The situation is reminiscent of vitamin A and its precursor beta-carotene in that the active form of vitamin A, all-*trans*-retinoic acid, is strongly antiproliferative in vitro, cancer-therapeutic and cancer-preventative in defined clinical settings, and its ingestion is epidemiologically linked to decreased cancer risk; but controlled trials of dietary supplementation have shown no benefit on cancer incidence or mortality [38, 44, 66].

Moreover, recent studies highlight the necessity of interpreting these supportive ecological and epidemiological data with caution. In a recent large longitudinal, nested, case-controlled study involving 622 prostate cancer cases and 1,451 matched controls, a U-shaped curve for prostate cancer risk and serum 25-OH vitamin D levels was found – a higher risk of prostate cancer with low (less than or equal to 19 nmol/l) and high (greater than or equal to 80 nmol/l) serum 25-OH vitamin D serum concentrations [106]. Earlier epidemiological studies did not find this U-shaped risk because of their smaller number of subjects and smaller variation in vitamin D serum concentrations. Moreover, by comparing high versus low serum 25-OH levels and not serum levels to the normal average concentration (40–60 nmol/l), any association found in these smaller studies would depend on where the serum 25-OH vitamin D levels of their homogeneous population were centered on this U-shaped curve. Finally, Tuohimaa et al. found

8

that the lag time between serum sampling and diagnosis considerably affected the relationship between low serum 25-OH level and prostate risk. Based on these results, earlier studies with lag times less than 10 years would find higher risks with low serum 25-OH vitamin D levels than those studies with lag times of 10 years or more. This suggests that vitamin D levels would not prevent prostate cancer, but would slow progression or mutation accumulation [106].

Similarly, not all observational studies have shown a consistent relationship between vitamin D and colorectal cancer risk. Recently, a randomized, multicenter, placebo-controlled trial showed that calcium supplementation (1200 mg elemental calcium) reduced colorectal adenomas in patients with serum 25-OH vitamin D levels >72 nmol/l (29.1 ng/ml) but had no effect in patients with lower serum 25-OH vitamin D levels [41]. Conversely, serum 25-OH vitamin D levels were only associated with reduced risk of colorectal adenomas in patients randomly assigned to receive calcium supplements. The authors interpreted this large study to demonstrate that calcium and vitamin D work together, not separately, to prevent colorectal carcinogenesis [41].

Despite superior methodology, the studies of Grau et al. [41] and Touhimaa et al. [106] clearly demonstrate the pitfalls of trying to use existing data on 25-OH vitamin D levels to make broad recommendations to the public on adequate vitamin D levels. Supplementing vitamin D to prevent deficiency as defined for calcium homeostasis (>25 nmol/l) is sufficient to decrease the prostate cancer risk but supplementing to >80 nmol/l increases the prostate cancer risk [106], while serum 25-OH vitamin D close to 80 nmol/l (>72 nmol/l) is required for calcium's beneficial effects on colorectal carcinogenesis [41]. Ultimately, only randomized, clinical trials of vitamin D supplementation will demonstrate the safety and efficacy of the suggested benefits of redefining sufficient versus insufficient/deficient levels of vitamin D to incorporate 25-OH vitamin D levels required for non-calcium homeostasis actions. Tamimi et al. concluded there was a "lack of clinical trial data demonstrating efficacy and safety of vitamin D supplementation" in cancer chemoprevention [102]. In 2003,

Trivedi et al. found no significant effects of vitamin D on total mortality or incidence of cancer or cardiovascular disease in a prospective, randomized double blind study of 100,000 IU oral vitamin D_3 every 4 months (approximately equivalent to 800 IU a day) for 5 years [105]. Finally, in a 2004 review, Vieth states "… there is no evidence from randomized clinical trials showing efficacy of vitamin D (cholecalciferol) for anything beyond osteoporosis …" [111].

Taken together, the negative results from a 5-year, prospective, double-blind, placebo-controlled clinical trial [105] and the presence of recent, additional data from the above-described larger observational studies, contradict the claim that a benefit to overall mortality or incidence of epithelial-derived cancers will be derived by supplementing vitamin D to any level above clear-cut deficiency as defined for skeletal health.

8.7 Diet/Supplementation Versus Sunshine

8.7.1 The Case for Diet

Overwhelming evidence, including nearly all the references cited in this chapter, asserts that vitamin D through dietary supplementation can correct both deficiency and insufficiency (except for those patients with gastrointestinal/malabsorption causes of vitamin D deficiency). From within the endocrine community, consensus groups, editorials, reviews, and letters, recognizing the carcinogenic potential of UV radiation in sunshine, call for more dietary vitamin D supplementation and not more sunshine (for example see Refs. 9, 11, 82, 107, 110, 111). Dietary supplementation or intramuscular (IM) injection is so efficacious and efficient that advocates for increasing sun exposure also recommend these routes as a method for vitamin D repletion (for example see Refs. 48 and 52).

Despite this consensus on the efficacy and efficiency of dietary supplementation, two primary arguments against dietary supplementation exist. The first major criticism of dietary vitamin D is that given the currently limited number of foods that are fortified, adequate amounts of vitamin from the standard American diet are difficult to obtain. The current recommended daily intakes of vitamin D interpret adequate as preventing vitamin D deficiency (i.e., maintaining serum 25-OH vitamin D levels above 25 nmol/l) [110] and recommend 200 IU a day for children and younger adults (<50 years old), 400 IU a day for older adults (>50 years old), and 600–800 IU a day for the elderly (>70 years old) [59]. In this regard, eight ounces of fortified milk or orange juice contains 100 IU (2.5 µg), the amount of vitamin D found in approximately half a teaspoon of cod-liver oil [50, 110]. Beyond these two foods, at least in the United States (margarine is supplemented in some parts of Europe), the population would have to rely on salmon, mackerel, and sardines [55, 69]. Moreover, foods that claim to be fortified, such as milk, have been shown in the early 1990s to have variable and sometimes inadequate or excessive amounts of vitamin D [16, 56], although this has not recently been reassessed.

However, this scarcity argument fails to acknowledge the contribution of incidental cutaneous sun exposure to total 25-OH vitamin D stores. In one study, a suberythemogenic exposure of 5% of the skin surface produced mean serum 25-OH vitamin D levels of 35 nmol/l, already above the deficient range [19]. These data were produced in elderly patients and so would underestimate the serum 25-OH levels from the same UVB exposure to younger populations. The surface area of the face and the back of the hands represents >5% of the total body surface area (BSA), and it is likely that despite the best efforts of sun protection campaigns, most non-institutionalized or non-home-bound people at baseline already receive at least 15 minutes of incidental unprotected sun exposure daily to these areas without further encouraging additional unprotected sun exposure [73].

Thus, incorporating incidental sun exposure mandates a dietary intake of only 5 µg (200 IU) per day to ensure the virtual absence of deficiency [110]. Furthermore, daily dietary vitamin D intake to exceed the highest current definition of insufficiency (<50 nmol/l) would then be 15 µg (600 IU) per day – within the realm of current dietary fortification. The demand for 1,000 IU per day from diet applies to distinct subsets of the population, such as those who live in submarines and the frail elderly with mobility issues limiting their access to incidental sunshine [110]. In these subgroups, the 25-OH vitamin D levels can be easily and sufficiently raised out of the insufficient and deficient ranges with as little as two 50,000 IU vitamin pills every 4 months [105].

The second criticism of dietary vitamin D supplementation is that excessive dietary intake of vitamin D can lead to vitamin D toxicity [1], while cutaneous production of vitamin D alone does not [45]. Recent reviews have summarized an impressive amount of data showing that hypervitaminosis D from diet is more a theoretical concern than a reality [11, 110]. Although controversial, serum 25-OH vitamin D levels as high as 140 nmol/l are considered safe [1, 110]. As stated above, maximal supplemental daily dietary intake of 1,000 IU a day would prevent deficiency and insufficiency by any current definition. As serum 25-OH levels reflect the summation from diet and cutaneous exposure, the addition of daily sun exposure to maximal dietary consumption could then lead to serum 25-OH levels exceeding 140 nmol/l.

Vieth estimated that keeping total cumulative daily vitamin D intake (i.e., from diet and UV radiation) to less than 40,000 IU a day will maintain serum 25-OH vitamin D levels <220 nmol/l, preventing frank hypervitaminosis D (i.e., hypercalcemia) [110]. Moreover, limiting total cumulative daily vitamin D to less than 10,000 IU a day will maintain serum 25-OH vitamin D levels <140 nmol/l, preventing hypercalciuria – an early sign of hypervitaminosis D [110]. The maximal estimate for cutaneous production of vitamin D assumes that one sunburn over 90% of BSA produces 20,000 IU of cutaneous vitamin D [110]. Then, with a daily dietary intake of 1000 IU, the most unlikely scenario of daily sunburns to >45% of total BSA would exceed cumulative daily intake of 10,000 IU a day. Thus, keeping dietary supplementation to levels of 1000 IU or lower would avoid the complications of the early signs of hypervitaminosis D (i.e., hypercalciuria and loss of bone mineral density) and still afford the clinical benefits seen in some randomized con-

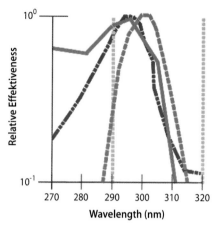

Fig. 8.5. Cutaneous vitamin D synthesis cannot be dissociated from the harmful effects of UV radiation. The *dashed-dotted line (blue)* shows the spectrum of previtamin D_3 formation obtained from plotting the reciprocal of photoenergy ($1/W\ cm^{-2}$) (adapted from Ref. 74) by converting from a linear to power-of-10 scale). The *dashed line (green)* represents both the action spectrum of the induction of squamous cell carcinoma in humans mathematically derived from experimental data obtained from murine skin, and the wavelength dependence of the induction of DNA damage, in this case cyclobutane pyrimidine dimers (CPD), in human skin (adapted from Ref. 24). The *solid line (red)* shows the erythema action spectrum from human skin (m^2/J) (adapted from Ref. 90). Note that the peaks of these three curves all occur within the UVB spectrum (290–320 nm; indicated by the *dashed gray line*)

trolled studies, especially if given with concurrent calcium supplementation.

8.7.2 The Case Against Sunshine

On the other hand, there are many reasons why cutaneous production is a suboptimal way to obtain vitamin D. First, the peaks of the action spectrum curves for previtamin D_3 formation, cutaneous erythema induction, and formation of cyclobutane pyrimidine dimers in DNA, all overlap within the range of UVB (290–320 nm) and cannot be dissociated (Fig. 8.5) [24, 74, 90]. UVB and UVA in natural sunlight are both well-documented carcinogens, contributing to the risk of basal cell carcinoma, squamous cell carcinoma, and malignant melanoma [24, 35,

91]. Conversely, it has been proposed that vitamin D metabolites mediate the putative mortality and antitumorigenic effects of increased UV radiation exposure [40], although to date this has not been validated by randomized, prospective clinical trials in humans [105]. In spite of this, in this case, dietary supplementation of vitamin D fortuitously allows patients to obtain the well-documented cancer-preventative effect of avoiding UV exposure without foregoing the putative chemopreventive and/or chemotherapeutic effects of UV radiation (i.e., cutaneous vitamin D production).

Second, in human in vivo clinical trials, single agents are usually only partially chemopreventive, such as for colon neoplasms, while combination therapies, such as for leukemias, are highly chemopreventive [65]. Further, the data suggest that vitamin D supplementation requires concurrent calcium supplementation for efficacy, such as for preventing colon polyps and falls in the elderly [8, 41]. The combination of calcium and vitamin D is readily available in one pill, and calcium-rich foods can be fortified with vitamin D. In contrast, sunlight-driven cutaneous production of vitamin D does not provide calcium.

Third, sunshine alone to obtain suggested levels of vitamin D is inefficient and cumbersome versus dietary supplementation. Based on "maximal" estimates of cutaneous vitamin D production (approximately 7,800–20,000 IU from exposing 90% of BSA to one minimal erythema dose, 1 MED), Holick recommended exposure of hands, face and arms (27% BSA), or arms and legs (54% BSA), for a period equal to 25% of the time required to obtain 1 MED [19, 49, 52]. Subsequent to this recommended exposure, patients should then either withdraw from sunlight or apply sun protection methods. Unfortunately, it is easy to calculate incorrectly the exposure time or forget to withdraw from unprotected sun exposure. As few as six sunburns in a lifetime have been shown to increase an individual's risk of non-melanoma and melanoma skin cancer [57, 60, 91, 108]. Alternatively, as few as one or two 50,000 IU pills a month could achieve this same monthly vitamin D intake with no cutaneous exposure.

Fourth, cutaneous vitamin D_3 production is compromised in all groups at high risk of vitamin D deficiency, as described in the following sections.

8.7.2.1 The Elderly

The elderly, especially those over 70 years old, are particularly at risk because they often have decreased 7-DHC in the skin due to decreased skin thickness, decreased mobility or institutionalization that minimizes sun exposure, decreased renal production of $1,25-(OH)_2$ vitamin D, and decreased intake of fortified foods [6, 83]. Accordingly, the increase in serum 25-OH vitamin D levels in the elderly from planned unprotected exposure to sunlight or artificial UV radiation [17, 55, 95] was less than that achieved by dietary or IM supplementation in the osteoporosis fracture-prevention studies (Table 8.4).

8.7.2.2 Darkly Pigmented Racial Groups

Dark skin pigmentation, especially in African–Americans, is correlated with increased risk of vitamin D deficiency. Melanin in the skin competes with 7-DHC for absorption of UVB photons. Pigmented and non-pigmented skin have the same overall capacity to make vitamin D_3, but pigmented skin requires longer exposure times [18]. As a group, dark-skinned individuals are likely to be lactose-intolerant and hence to avoid milk, a major dietary source of vitamin D. One study found that that African–American women aged 15–49 years have a more than tenfold higher risk of being vitamin D deficient (serum 25-OH vitamin D <25 nmol/l) versus white counterparts (12.2% vs 0.5% of the population) [84]. In addition, the re-emergence of rickets in American infants is virtually restricted to dark-skinned infants exclusively fed human milk, a poor source of dietary vitamin D [64, 114]. However, drawing on the well-documented success of vitamin D dietary supplementation in eradicating rickets in the last century, the American Academy of Pediatrics recommends for at-risk infants and children dietary supple-

mentation of vitamin D, not sun exposure with its undisputed long-term risks [34, 114].

8.7.2.3 Those Living in Certain Environments

During wintertime, at latitudes above 35°, for example New York City (40°), Paris (48°), and Moscow (55°), sunlight is incapable of producing previtamin D_3 [52]. Independent of latitude, air pollution, usually associated with urban settings, also reduces the amount of UVB reaching the earth's surface [3]. In spite of this, latitude alone will not predict vitamin D deficiency as it fails to account for the contribution of dietary vitamin D to overall vitamin D stores. For example, Van der Wielen et al. found that elderly Europeans living in Northern Europe had 25-OH vitamin D levels (mean 54 nmol/l) well above deficient and insufficient (by most definitions) levels as a consequence primarily of their diet high in vitamin D-containing fish oils and/or vitamin D supplementation [109].

At the end of winter in Boston (latitude 42°), where very little to no previtamin D_3 can be made from sunlight during winter, the mean vitamin D level (50 nmol/l) was well above deficient or insufficient levels for four different age groups (18–29, 30–39, 40–49, and 50+ years) [103]. Moreover, in all but the youngest age group, there was no statistically significant difference in vitamin D levels between summer and winter, showing that 25-OH vitamin D stores are only slowly depleted during obligate periods of low production of previtamin D. The findings also imply that at least for this Boston population, diet and/or supplementation may be important contributors to preventing vitamin D deficiency and insufficiency. Finally, these authors found that a 400 IU multivitamin pill easily corrected their definition of insufficiency/deficiency in the 18–29 year group [103]. Importantly, winter dietary supplementation thus avoids the need for carcinogenic UVB exposure from artificial sources, such as tanning beds, to maintain sufficient 25-OH vitamin D in this 18–29 year age group who have decades to develop the harmful consequences of chronic, unprotected UV exposure [60, 115, 117].

8.7.2.4 Cultural and Lifestyle Practices

Cultural and lifestyle practices that actively or inadvertently minimize unprotected sun exposure and/or the use of sun protection and sunscreens also have the potential to adversely affect cutaneous vitamin D production. One study found that Southern Europeans, where latitude favors cutaneous previtamin D synthesis, nevertheless often have insufficient winter time levels of 25-OH vitamin D, attributed to sun avoidance, long clothing, and increased disability in the absence of effective dietary supplementation [109]. However, while lifestyle/cultural practices may reduce cutaneous production of vitamin D even more so than latitude, dietary supplementation can prevent ensuing vitamin D deficiency. Thus, Glerup et al. showed that in veiled Arab females, although they continued to have minimal cutaneous production of vitamin D, vitamin D injections or oral supplementation efficiently corrected the overall state of vitamin D deficiency [36].

8.7.2.5 Sunscreen Use

Evidence shows that effective sunscreen use does not cause vitamin D deficiency or insufficiency in otherwise healthy individuals. However, two papers are extensively cited in support of the opposite conclusion [74, 75]. Matsuoka et al. documented that in a laboratory situation, applying ethanol with 5% (wt/vol) p-aminobenzoic acid (PABA) to pieces of normal human skin prevented UVB-induced conversion of 7-DHC to previtamin D_3 [74]. Matsuoka et al. also showed that applying a 5% PABA solution, having a sun protection factor (SPF) of 8, 1 hour before 1 MED UVB exposure from a light box effectively prevented an increase in serum vitamin D_3 levels in four patients [52, 74]. These findings are not surprising given the great overlap of the absorption spectrum for previtamin D_3 formation (Fig. 8.5) and of PABA sunscreen (260–340 nm, peak approximately 310 nm) [67].

Most people's real life experience with sunscreen is that despite its application, they still tan or sunburn after prolonged sun exposure.

SPF is a strictly defined and FDA-regulated measurement based on applying 2 mg/cm^2 of sunscreen [67]. Studies have shown that in reality, most users apply insufficient amounts of sunscreen to meet this FDA standard, and the true SPF obtained is usually <50% of that written on the package [5]. In addition, sunscreen efficacy depends on its uniform application to all relevant body parts, its durability, its degree of "water-resistance", and its reapplication after it gets washed off [67]. Finally, even if properly applied, by definition sunscreens do not provide complete blockage of UVB: SPF 15 allows approximately 6% of UVB photons to penetrate the skin [67]. Thus, it is highly unlikely that even regular sunscreen use will cause vitamin D deficiency or insufficiency. Indeed, Matsuoka et al. also reported a case-controlled study of 20 "long-term" sunscreen users versus 20 controls [75]. Long-term users were patients with documented skin cancer who stated they had regularly used sunscreen on all exposed skin for longer than 1 year at their doctor's recommendation. In these users, the mean 25-OH vitamin D level was 40.2±3.2 nmol/l versus 91±6.2 nmol/l in controls. Thus, although they had lower 25-OH vitamin D levels than controls, long-term users were not vitamin D deficient, contradicting the claims that sunscreen use causes vitamin D deficiency (mean 25-vitamin OH levels <20 nmol/l) or insufficiency (mean 25-vitamin OH levels 20–37.5 nmol/l) by widely used definitions [37, 61, 70, 104]. Thus, this study shows that UVB transmission through sunscreen as regularly used by long-term users, with diet-derived 25-OH vitamin D (not quantified by this study), was sufficient to maintain non-deficient serum 25-OH vitamin D levels.

In fact, Marks et al. demonstrated in a prospective randomized controlled clinical trial that use of SPF 17 sunscreen at levels sufficient to prevent actinic keratoses did not induce vitamin D insufficiency or deficiency [73]. This trial of 113 subjects took place in Australia at a latitude of 37° (a location where the angle of the sun is not sufficient to produce previtamin D_3 all year round), and both sunscreen use and sun exposure were quantified for each individual. Moreover, all subjects were instructed to avoid sun exposure around the middle of the day and to wear hats and other appropriate clothing for sun protection. None of the subjects in any of the groups, even those 70 years and older, was vitamin D deficient or insufficient (<50 nmol/l) at baseline, and after the intervention, 25-OH vitamin D levels rose by greater than 7 nmol/l in all groups with no statistically significant differences between the mean changes for sunscreen users and controls in any category [73].

Other studies support the conclusion of Marks et al. that, in real life, sunscreen use does not cause vitamin D deficiency [73]. One prospective, longitudinal clinical trial comprising 24 subjects studied for 2 years in Barcelona (latitude 41° North) found no vitamin D deficiency or secondary hyperparathyroidism in sunscreen users (SPF 15), and no statistically significant change in markers of bone remodeling in sunscreen users versus controls [26]. Finally, another cross-sectional study of eight patients with xeroderma pigmentosum who practiced rigorous photoprotection found that the serum 25-OH vitamin D levels were in the low normal range (mean 44.5 ± 3.75 nmol/l with normal levels of PTH and calcitriol after 6 years of follow-up [99].

In summary, an evaluation of the available literature shows that even when used effectively to reduce actinic damage, sunscreens do not cause vitamin D deficiency or adversely affect markers of bone remodeling and skeletal homeostasis.

8.8 Conclusion

Given the scarcity of naturally occurring dietary vitamin D, humans may once have depended on unprotected exposure to natural sunlight as the primary environmental source of vitamin D, at least during those periods of the year when sunlight can produce previtamin D_3 in the skin [111]. However, chronic unprotected exposure to carcinogenic UV radiation in sunlight not only subjects individuals to the adverse effects of photoaging, but also greatly increases their risk of developing skin cancer. This risk is further exacerbated by the extended life-span of humans in the 21st century. Fortunately, nature has provided humans with a non-carcinogenic

alternative – intestinal absorption of vitamin D-fortified foods and/or dietary supplements.

This issue of dietary absorption versus cutaneous production to maximize vitamin D intake becomes particularly relevant in light of the proposed upward revision of recommendations for optimal serum 25-OH vitamin D levels. Currently, these higher recommended serum 25-OH vitamin D levels are thought to maximize skeletal health, muscle strength and balance in the elderly and to offer other as-yet-unproven benefits, such as lowering blood pressure, reduction in mortality from some cancers, and prevention of autoimmune diseases, such as type I diabetes and multiple sclerosis. Accepting the above as possible, even if unproven in most instances, argues in favor of increasing the recommended daily intake of vitamin D to achieve these higher serum 25-OH vitamin D levels.

In this regard, augmenting cutaneous production of previtamin D_3 through intentional unprotected exposure to UV radiation is still proposed today by some as the preferred method to maintain serum 25-OH vitamin D levels at or above a desired level. That is, some recommend a carcinogen in moderation despite an equally effective non-carcinogenic alternative. This recommendation is doubly problematic in that cutaneous production does not offer concomitant calcium supplementation, and is known to be inefficient in those statistically most "at risk" of having low serum 25-OH levels, such as the elderly and dark-skinned persons. Moreover, the "tanning message" inherent in this proposal targets primarily young, lighter-skinned individuals who have the longest available time to develop the harmful consequences of chronic, unprotected UV exposure [60, 115, 117] and who are most likely to be achieving maximal potential previtamin D photosynthesis through incidental sun exposure, even when wearing sunscreen.

Rather than promote unprotected chronic sun exposure in younger, lighter-skinned individuals, desirable serum 25-OH vitamin D levels in this population would seem easy to maintain by incidental protected sun exposure and customary diet. Vieth estimated that 200 IU per day as a sole source of vitamin D may be sufficient to prevent skeletal signs of vitamin D deficiency (i.e., maintain serum 25-OH at 25 nmol/l) [110]. Thus, daily intake of two eight-ounce glasses of fortified milk or orange juice or one vitamin pill or incidental protected exposure of the face and backs of hands to 0.25% MED of UVB radiation, would each achieve adequate serum 25-OH levels to prevent vitamin D deficiency. In spite of this, conservatively assuming a baseline vitamin D intake of only one of these three scenarios, additional dietary vitamin D intake of 400 IU, which has been shown to increase serum 25-OH vitamin D by approximately 45 nmol/l [110], to 600 IU a day readily prevents vitamin D insufficiency by most standards (<50 nmol/l). Finally, for those groups at high statistical risk of vitamin D deficiency, concurrent dietary calcium and vitamin D (800–1,000 IU a day) supplementation maintains sufficient vitamin D and calcium stores, free of the deleterious effects of carcinogenic UV radiation.

References

1. Adams JS, Lee G (1997) Gains in bone mineral density with resolution of vitamin D intoxication. Ann Intern Med 127:203–206
2. Adams JS, Clemens TL, Parrish JA, Holick MF (1982) Vitamin-D synthesis and metabolism after ultraviolet irradiation of normal and vitamin-D-deficient subjects. N Engl J Med 306:722–725
3. Agarwal KS, Mughal MZ, Upadhyay P, Berry JL, Mawer EB, Puliyel JM (2002) The impact of atmospheric pollution on vitamin D status of infants and toddlers in Delhi, India. Arch Dis Child 87:111–113
4. Albert MR, Ostheimer KG (2003) The evolution of current medical and popular attitudes toward ultraviolet light exposure: Part 3. J Am Acad Dermatol 49:1096–1106
5. Bech-Thomsen N, Wulf HC (1992) Sunbathers' application of sunscreen is probably inadequate to obtain the sun protection factor assigned to the preparation. Photodermatol Photoimmunol Photomed 9:242–244
6. Bell NH (1995) Vitamin D metabolism, aging, and bone loss. J Clin Endocrinol Metab 80:1051
7. Bischoff-Ferrari HA, Borchers M, Gudat F, Durmuller U, Stahelin HB, Dick W (2004) Vitamin D receptor expression in human muscle tissue decreases with age. J Bone Miner Res 19:265–269
8. Bischoff-Ferrari HA, Dawson-Hughes B, Willett WC, Staehelin HB, Bazemore MG, Zee RY, Wong JB (2004) Effect of vitamin D on falls: a meta-analysis. JAMA 291:1999–2006
9. Boucher BJ (2001) Sunlight "D"ilemma. Lancet 357:961
10. Bringhurst RF, Demay MB, Kronenberg HM (2003) Hormones and disorders of mineral metabolism. In: Larsen PR, Kronenberg HM, Melmed S, Polonsky KS (eds) Williams textbook of endocrinology. Saunders, Philadelphia, pp 1317–1320
11. Calvo MS, Whiting SJ, Barton CN (2005) Vitamin D intake: a global perspective of current status. J Nutr 135:310–316
12. Chapuy MC, Meunier PJ (1996) Prevention of secondary hyperparathyroidism and hip fracture in elderly women with calcium and vitamin D3 supplements. Osteoporos Int 6 [Suppl 3]:60–63
13. Chapuy MC, Arlot ME, Duboeuf F, Brun J, Crouzet B, Arnaud S, Delmas PD, Meunier PJ (1992) Vitamin D3 and calcium to prevent hip fractures in the elderly women. N Engl J Med 327:1637–1642
14. Chapuy MC, Arlot ME, Delmas PD, Meunier PJ (1994) Effect of calcium and cholecalciferol treatment for three years on hip fractures in elderly women. BMJ 308:1081–1082
15. Chapuy MC, Pamphile R, Paris E, Kempf C, Schlichting M, Arnaud S, Garnero P, Meunier PJ (2002) Combined calcium and vitamin D3 supplementation in elderly women: confirmation of reversal of secondary hyperparathyroidism and hip fracture risk: The Decalyos II study. Osteoporos Int 13:257–264
16. Chen TC, Shao A, Heath H 3rd, Holick MF (1993) An update on the vitamin D content of fortified milk from the United States and Canada. N Engl J Med 329:1507
17. Chuck A, Todd J, Diffey B (2001) Subliminal ultraviolet-B irradiation for the prevention of vitamin D deficiency in the elderly: a feasibility study. Photodermatol Photoimmunol Photomed 17:168–171
18. Clemens TL, Adams JS, Henderson SL, Holick MF (1982) Increased skin pigment reduces the capacity of skin to synthesise vitamin D3. Lancet 1:74–76
19. Davie MW, Lawson DE, Emberson C, Barnes JL, Roberts GE, Barnes ND (1982) Vitamin D from skin: contribution to vitamin D status compared with oral vitamin D in normal and anticonvulsant-treated subjects. Clin Sci (Lond) 63:461–472
20. Dawson-Hughes B, Dallal GE, Krall EA, Harris S, Sokoll LJ, Falconer G (1991) Effect of vitamin D supplementation on wintertime and overall bone loss in healthy postmenopausal women. Ann Intern Med 115:505–512
21. Dawson-Hughes B, Harris SS, Krall EA, Dallal GE, Falconer G, Green CL (1995) Rates of bone loss in postmenopausal women randomly assigned to one of two dosages of vitamin D. Am J Clin Nutr 61:1140–1145

22. Dawson-Hughes B, Harris SS, Dallal GE (1997) Plasma calcidiol, season, and serum parathyroid hormone concentrations in healthy elderly men and women. Am J Clin Nutr 65:67–71

23. Dawson-Hughes B, Harris SS, Krall EA, Dallal GE (1997) Effect of calcium and vitamin D supplementation on bone density in men and women 65 years of age or older. N Engl J Med 337:670–676

24. de Gruijl FR (1999) Skin cancer and solar UV radiation. Eur J Cancer 35:2003–2009

25. EURODIAB Substudy 2 Study Group (1999) Vitamin D supplement in early childhood and risk for Type I (insulin-dependent) diabetes mellitus. Diabetologia 42:51–54

26. Farrerons J, Barnadas M, Rodriguez J, Renau A, Yoldi B, Lopez-Navidad A, Moragas J (1998) Clinically prescribed sunscreen (sun protection factor 15) does not decrease serum vitamin D concentration sufficiently either to induce changes in parathyroid function or in metabolic markers. Br J Dermatol 139:422–427

27. Feskanich D, Willett WC, Colditz GA (2003) Calcium, vitamin D, milk consumption, and hip fractures: a prospective study among postmenopausal women. Am J Clin Nutr 77:504–511

28. Fry A, Verne J (2003) Preventing skin cancer. BMJ 326:114–115

29. Gallagher JC (2004) The effects of calcitriol on falls and fractures and physical performance tests. J Steroid Biochem Mol Biol 89-90:497–501

30. Garland CF (2003) More on preventing skin cancer: sun avoidance will increase incidence of cancers overall. BMJ 327:1228

31. Garland CF (2003) Sun avoidance will increase overall cancer incidence. http://bmj.com/cgi/eletters/326/7381/114#28894

32. Garland CF, Garland FC, Gorham ED (1991) Can colon cancer incidence and death rates be reduced with calcium and vitamin D? Am J Clin Nutr 54:193S–201S

33. Garland CF, Garland FC, Gorham ED (1999) Calcium and vitamin D. Their potential roles in colon and breast cancer prevention. Ann N Y Acad Sci 889:107–119

34. Gartner LM, Greer FR (2003) Prevention of rickets and vitamin D deficiency: new guidelines for vitamin D intake. Pediatrics 111:908–910

35. Gilchrest BA (1993) Sunscreens – a public health opportunity. N Engl J Med 329:1193–1194

36. Glerup H, Mikkelsen K, Poulsen L, Hass E, Overbeck S, Andersen H, Charles P, Eriksen EF (2000) Hypovitaminosis D myopathy without biochemical signs of osteomalacic bone involvement. Calcif Tissue Int 66:419–424

37. Gloth FM 3rd, Gundberg CM, Hollis BW, Haddad JG Jr, Tobin JD (1995) Vitamin D deficiency in homebound elderly persons. JAMA 274:1683–1686

38. Goodman GE, Thornquist MD, Balmes J, Cullen MR, Meyskens FL Jr, Omenn GS, Valanis B, Williams JH Jr (2004) The beta-carotene and retinol efficacy trial: incidence of lung cancer and cardiovascular disease mortality during 6-year follow-up after stopping beta-carotene and retinol supplements. J Natl Cancer Inst 96:1743–1750

39. Grant WB (2003) The benefits of sunlight outweigh the harms. http://bmj.com/cgi/eletters/326/7381/114#28894

40. Grant WB, Garland CF (2002) Evidence supporting the role of vitamin D in reducing the risk of cancer. J Intern Med 252:178–179; author reply 179–180

41. Grau MV, Baron JA, Sandler RS, Haile RW, Beach ML, Church TR, Heber D (2003) Vitamin D, calcium supplementation, and colorectal adenomas: results of a randomized trial. J Natl Cancer Inst 95:1765–1771

42. Heaney RP (1999) Lessons for nutritional science from vitamin D. Am J Clin Nutr 69:825–826

43. Heikinheimo RJ, Inkovaara JA, Harju EJ, Haavisto MV, Kaarela RH, Kataja JM, Kokko AM, Kolho LA, Rajala SA (1992) Annual injection of vitamin D and fractures of aged bones. Calcif Tissue Int 51:105–110

44. Hennekens CH, Buring JE, Manson JE, Stampfer M, Rosner B, Cook NR, Belanger C, LaMotte F, Gaziano JM, Ridker PM, Willett W, Peto R (1996) Lack of effect of long-term supplementation with beta carotene on the incidence of malignant neoplasms and cardiovascular disease. N Engl J Med 334:1145–1149

45. Holick MF (1981) The cutaneous photosynthesis of previtamin D3: a unique photoendocrine system. J Invest Dermatol 77:51–58

46. Holick MF (1990) The use and interpretation of assays for vitamin D and its metabolites. J Nutr 120 [Suppl 11]:1464–1469

47. Holick MF (1999) Vitamin D: photobiology, metabolism, mechanism of action, and clinical applications. In: Favus M (ed) Primer on the metabolic bone diseases and disorders of mineral metabolism. Lippincott Williams & Wilkins, Philadelphia, pp 92–98

48. Holick MF (2001) Sunlight "D"ilemma: risk of skin cancer or bone disease and muscle weakness. Lancet 357:4–6

49. Holick MF (2002) Vitamin D: the underappreciated D-lightful hormone that is important for skeletal and cellular health. Curr Opin Endocrinol Diabetes 9:87–98

50. Holick MF (2004) In reply to "the physiology and treatment of vitamin D deficiency". Mayo Clin Proc 79:694–695

51. Holick MF (2004) In reply to "vitamin D deficiency and chronic pain: cause and effect or epiphenomenon?". Mayo Clin Proc 79:699

8

52. Holick MF (2004) Vitamin D: importance in the prevention of cancers, type 1 diabetes, heart disease, and osteoporosis. Am J Clin Nutr 79:362–371

53. Holick M, Jenkins M (2004) The UV advantage. ibooks, New York

54. Holick MF, MacLaughlin JA, Clark MB, Holick SA, Potts JT Jr, Anderson RR, Blank IH, Parrish JA, Elias P (1980) Photosynthesis of previtamin D3 in human skin and the physiologic consequences. Science 210:203–205

55. Holick MF, Matsuoka LY, Wortsman J (1989) Age, vitamin D, and solar ultraviolet. Lancet 2:1104–1105

56. Holick MF, Shao Q, Liu WW, Chen TC (1992) The vitamin D content of fortified milk and infant formula. N Engl J Med 326:1178–1181

57. Hunter DJ, Colditz GA, Stampfer MJ, Rosner B, Willett WC, Speizer FE (1990) Risk factors for basal cell carcinoma in a prospective cohort of women. Ann Epidemiol 1:13–23

58. Hypponen E, Laara E, Reunanen A, Jarvelin MR, Virtanen SM (2001) Intake of vitamin D and risk of type 1 diabetes: a birth-cohort study. Lancet 358:1500–1503

59. Institute of Medicine (1997) Dietary reference intakes for calcium, phosphorus, magnesium, vitamin D, and fluoride. Vitamin D. Food, and Nutrition Board, Standing Committee on the Scientific Evaluation of Dietary Reference Intakes. National Academy Press, Washington DC

60. Kennedy C, Bajdik CD, Willemze R, De Gruijl FR, Bouwes Bavinck JN; Leiden Skin Cancer Study (2003) The influence of painful sunburns and lifetime sun exposure on the risk of actinic keratoses, seborrheic warts, melanocytic nevi, atypical nevi, and skin cancer. J Invest Dermatol 120:1087–1093

61. Kinyamu HK, Gallagher JC, Rafferty KA, Balhorn KE (1998) Dietary calcium and vitamin D intake in elderly women: effect on serum parathyroid hormone and vitamin D metabolites. Am J Clin Nutr 67:342–348

62. Klein G (1999) Nutritional rickets and osteomalacia. In: Flavus M (ed) Primer on the metabolic bone diseases and disorders of mineral metabolism. Lippincott Williams & Wilkins, Philadelphia, pp 315–319

63. Krause R, Buhring M, Hopfenmuller W, Holick MF, Sharma AM (1998) Ultraviolet B and blood pressure. Lancet 352:709–710

64. Kreiter SR, Schwartz RP, Kirkman HN Jr, Charlton PA, Calikoglu AS, Davenport ML (2000) Nutritional rickets in African American breast-fed infants. J Pediatr 137:153–157

65. Lamprecht SA, Lipkin M (2003) Chemoprevention of colon cancer by calcium, vitamin D and folate: molecular mechanisms. Nat Rev Cancer 3:601–614

66. Lee IM, Cook NR, Manson JE, Buring JE, Hennekens CH (1999) Beta-carotene supplementation and incidence of cancer and cardiovascular disease: the Women's Health Study. J Natl Cancer Inst 91:2102–2106

67. Levy S (2001) Sunscreens. In: Wolverton S (ed) Comprehensive dermatologic drug therapy. Saunders, Philadelphia, pp 632–646

68. Lips P, Wiersinga A, van Ginkel FC, Jongen MJ, Netelenbos JC, Hackeng WH, Delmas PD, van der Vijgh WJ (1988) The effect of vitamin D supplementation on vitamin D status and parathyroid function in elderly subjects. J Clin Endocrinol Metab 67:644–650

69. Lips P, Graafmans WC, Ooms ME, Bezemer PD, Bouter LM (1996) Vitamin D supplementation and fracture incidence in elderly persons. A randomized, placebo-controlled clinical trial. Ann Intern Med 124:400–406

70. MacLaughlin JA, Anderson RR, Holick MF (1982) Spectral character of sunlight modulates photosynthesis of previtamin D3 and its photoisomers in human skin. Science 216:1001–1003

71. Majewski S, Skopinska M, Marczak M, Szmurlo A, Bollag W, Jablonska S (1996) Vitamin D3 is a potent inhibitor of tumor cell-induced angiogenesis. J Investig Dermatol Symp Proc 1:97–101

72. Malabanan A, Veronikis IE, Holick MF (1998) Redefining vitamin D insufficiency. Lancet 351:805–806

73. Marks R, Foley PA, Jolley D, Knight KR, Harrison J, Thompson SC (1995) The effect of regular sunscreen use on vitamin D levels in an Australian population. Results of a randomized controlled trial. Arch Dermatol 131:415–421

74. Matsuoka LY, Ide L, Wortsman J, MacLaughlin JA, Holick MF (1987) Sunscreens suppress cutaneous vitamin D3 synthesis. J Clin Endocrinol Metab 64:1165–1168

75. Matsuoka LY, Wortsman J, Hanifan N, Holick MF (1988) Chronic sunscreen use decreases circulating concentrations of 25-hydroxyvitamin D. A preliminary study. Arch Dermatol 124:1802–1804

76. McMichael AJ, Hall AJ (2001) Multiple sclerosis and ultraviolet radiation: time to shed more light. Neuroepidemiology 20:165–167

77. Mehta RG, Mehta RR (2002) Vitamin D and cancer. J Nutr Biochem 13:252–264

78. Meyer HE, Smedshaug GB, Kvaavik E, Falch JA, Tverdal A, Pedersen JI (2002) Can vitamin D supplementation reduce the risk of fracture in the elderly? A randomized controlled trial. J Bone Miner Res 17:709–715

79. Michaelsson K, Melhus H, Bellocco R, Wolk A (2003) Dietary calcium and vitamin D intake in relation to osteoporotic fracture risk. Bone 32:694–703

80. Munger KL, Zhang SM, O'Reilly E, Hernan MA, Olek MJ, Willett WC, Ascherio A (2004) Vitamin D intake and incidence of multiple sclerosis. Neurology 62:60–65
81. Murayama A, Takeyama K, Kitanaka S, Kodera Y, Kawaguchi Y, Hosoya T, Kato S (1999) Positive and negative regulations of the renal 25-hydroxyvitamin D3 1alpha-hydroxylase gene by parathyroid hormone, calcitonin, and 1alpha,25-(OH)2D3 in intact animals. Endocrinology 140:2224–2231
82. NASA Occupational Health (2005) Position statement: risks and benefits of sun exposure. http://www.ohp.nasa.gov/topics/skin/risks_benefits_sun_exposure_mar05.pdf, 2005
83. Need AG, Morris HA, Horowitz M, Nordin C (1993) Effects of skin thickness, age, body fat, and sunlight on serum 25-hydroxyvitamin D. Am J Clin Nutr 58:882–885
84. Nesby-O'Dell S, Scanlon KS, Cogswell ME, Gillespie C, Hollis BW, Looker AC, Allen C, Doughertly C, Gunter EW, Bowman BA (2002) Hypovitaminosis D prevalence and determinants among African American and white women of reproductive age: third National Health and Nutrition Examination Survey, 1988–1994. Am J Clin Nutr 76:187–192
85. Norris JM (2001) Can the sunshine vitamin shed light on type 1 diabetes? Lancet 358:1476–1478
86. Okonofua F, Gill DS, Alabi ZO, Thomas M, Bell JL, Dandona P (1991) Rickets in Nigerian children: a consequence of calcium malnutrition. Metabolism 40:209–213
87. Ooms ME, Lips P, Roos JC, van der Vijgh WJ, Popp-Snijders C, Bezemer PD, Bouter LM (1995) Vitamin D status and sex hormone binding globulin: determinants of bone turnover and bone mineral density in elderly women. J Bone Miner Res 10:1177–1184
88. Ooms ME, Roos JC, Bezemer PD, van der Vijgh WJ, Bouter LM, Lips P (1995) Prevention of bone loss by vitamin D supplementation in elderly women: a randomized double-blind trial. J Clin Endocrinol Metab 80:1052–1058
89. Parfitt AM, Gallagher JC, Heaney RP, Johnston CC, Neer R, Whedon GD (1982) Vitamin D and bone health in the elderly. Am J Clin Nutr 36:1014–1031
90. Parrish JA, Jaenicke KF, Anderson RR (1982) Erythema and melanogenesis action spectra of normal human skin. Photochem Photobiol 36:187–191
91. Pfahlberg A, Kolmel KF, Gefeller O; Febim Study Group (2001) Timing of excessive ultraviolet radiation and melanoma: epidemiology does not support the existence of a critical period of high susceptibility to solar ultraviolet radiation-induced melanoma. Br J Dermatol 144:471–475
92. Pfeifer M, Begerow B, Minne HW (2002) Vitamin D and muscle function. Osteoporos Int 13:187–194
93. Plotnikoff GA, Quigley JM (2003) Prevalence of severe hypovitaminosis D in patients with persistent, nonspecific musculoskeletal pain. Mayo Clin Proc 78:1463–1470
94. Reichel H, Koeffler HP, Norman AW (1989) The role of the vitamin D endocrine system in health and disease. N Engl J Med 320:980–991
95. Reid IR, Gallagher DJ, Bosworth J (1986) Prophylaxis against vitamin D deficiency in the elderly by regular sunlight exposure. Age Ageing 15:35–40
96. Rigel DS (2002) Photoprotection: a 21st century perspective. Br J Dermatol 146 [Suppl 61]:34–37
97. Santmyire BR, Feldman SR, Fleischer AB Jr (2001) Lifestyle high-risk behaviors and demographics may predict the level of participation in sun-protection behaviors and skin cancer primary prevention in the united states: results of the 1998 National Health Interview Survey. Cancer 92:1315–1324
98. Shoback D, Marcus R, Bikle D, et al (2004) Metabolic bone disease. In: Greenspan FS, Gardner DG (eds) Basic and clinical endocrinology, 7th edn. Lange Medical Books/McGraw-Hill, New York, p 322
99. Sollitto RB, Kraemer KH, DiGiovanna JJ (1997) Normal vitamin D levels can be maintained despite rigorous photoprotection: six years' experience with xeroderma pigmentosum. J Am Acad Dermatol 37:942–947
100. Stene LC, Joner G (2003) Use of cod liver oil during the first year of life is associated with lower risk of childhood-onset type 1 diabetes: a large, population-based, case-control study. Am J Clin Nutr 78:1128–1134
101. Stumpf WE, Sar M, Reid FA, Tanaka Y, DeLuca HF (1979) Target cells for 1,25-dihydroxyvitamin D3 in intestinal tract, stomach, kidney, skin, pituitary, and parathyroid. Science 206:1188–1190
102. Tamimi RM, Lagiou P, Adami H-O, Trichopoulos D (2002) Comments on "evidence supporting the role of vitamin D in reducing the risk of cancer". J Intern Med 252:179–180
103. Tangpricha V, Pearce EN, Chen TC, Holick MF (2002) Vitamin D insufficiency among free-living healthy young adults. Am J Med 112:659–662
104. Thomas MK, Lloyd-Jones DM, Thadhani RI, Shaw AC, Deraska DJ, Kitch BT, Vamvakas EC, Dick IM, Prince RL, Finkelstein JS (1998) Hypovitaminosis D in medical inpatients. N Engl J Med 338:777–783
105. Trivedi DP, Doll R, Khaw KT (2003) Effect of four monthly oral vitamin D3 (cholecalciferol) supplementation on fractures and mortality in men and women living in the community: randomised double blind controlled trial. BMJ 326:469

106. Tuohimaa P, Tenkanen L, Ahonen M, Lumme S, Jellum E, Hallmans G, Stattin P, Harvei S, Hakulinen T, Luostarinen T, Dillner J, Lehtinen M, Hakama M (2004) Both high and low levels of blood vitamin D are associated with a higher prostate cancer risk: a longitudinal, nested case-control study in the Nordic countries. Int J Cancer 108:104–108

107. Utiger RD (1998) The need for more vitamin D. N Engl J Med 338:828–829

108. van Dam RM, Huang Z, Rimm EB, Weinstock MA, Spiegelman D, Colditz GA, Willett WC, Giovannucci E (1999) Risk factors for basal cell carcinoma of the skin in men: results from the health professionals follow-up study. Am J Epidemiol 150:459–468

109. van der Wielen RP, Lowik MR, van den Berg H, de Groot LC, Haller J, Moreiras O, van Staveren WA (1995) Serum vitamin D concentrations among elderly people in Europe. Lancet 346:207–210

110. Vieth R (1999) Vitamin D supplementation, 25-hydroxyvitamin D concentrations, and safety. Am J Clin Nutr 69:842–856

111. Vieth R (2004) Why the optimal requirement for vitamin D3 is probably much higher than what is officially recommended for adults. J Steroid Biochem Mol Biol 89-90:575–579

112. Vieth R (2004) Why "vitamin D" is not a hormone, and not a synonym for 1,25-dihydroxy-vitamin D, its analogs or deltanoids. J Steroid Biochem Mol Biol 89-90:571–573

113. Webb AR, Pilbeam C, Hanafin N, Holick MF (1990) An evaluation of the relative contributions of exposure to sunlight and of diet to the circulating concentrations of 25-hydroxyvitamin D in an elderly nursing home population in Boston. Am J Clin Nutr 51:1075–1081

114. Welch TR, Bergstrom WH, Tsang RC (2000) Vitamin D-deficient rickets: the reemergence of a once-conquered disease. J Pediatr 137:143–145

115. World Health Organization (2005) Sunbeds, tanning and UV exposure. http://www.who.int/mediacentre/factsheets/fs287/en/

116. Yaar M, Eller MS, Gilchrest BA (2002) Fifty years of skin aging. J Investig Dermatol Symp Proc 7:51–58

117. Young AR (2004) Tanning devices – fast track to skin cancer? Pigment Cell Res 17:2–9

Chapter 9

Modern Photoprotection of Human Skin

9

Jean Krutmann, Daniel Yarosh

Contents

9.1 Introduction

Prevention of premature skin aging is of constantly increasing importance to the general population. The mainstay of skin protective strategies is photoprotection, and as a practical consequence, UV filters are no longer used in sunscreens only, but are increasingly present at relevant concentrations in cosmetic products which are meant for daily use, such as foundations, make-ups, and creams and lotions. From a dermatological point of view, this development is positive since it greatly facilitates daily protection against the many harmful effects UV rays exert on human skin. In general, there is no doubt that UV filters are effective when properly used and dermatologists should encourage their use as the primary or basic strategy in photoprotection [22]. However, sunscreens are very frequently misapplied or underused, and this allows significant DNA damage to occur [10]. Further, there is increasing evidence that the efficacy of photoprotection provided by UV filters may differ for different biological endpoints such as prevention of erythema, protection of the skin immune system, prevention of premature skin aging or skin carcinogenesis. In other words, photoprotection provided by UV filters is not 100% perfect and thus, there is room for improvement.

Modern sunscreen products therefore frequently combine UV filters with one or more biologically active molecules ("actives") which provide photoprotection through mechanisms which are not based on absorption or reflection of UV rays and thus are completely different from UV filters. The protective capacity of actives relies instead on their capacity to prevent some of the biochemical and molecular consequences which occur in the skin after UV radiation has been absorbed. A popular example of such a secondary photoprotective strategy is the use of antioxidants in sunscreen products. During recent years many additional actives with quite diverse target points for photoprotection have been identified. In fact, the development of novel actives for skin photoprotection is a highly innovative and dynamic field, which in most instances has been directly stimulated and made possible by the most recent advances in cutaneous biological research [7]. It is therefore impossible to provide a complete overview of all actives that are currently being used in cosmetic products. Instead, some "prototypic molecules" representative of the most important and most frequently used groups of actives are discussed in this chapter to illustrate the relevance of these molecules as a secondary strategy for photoprotection. From these examples it will also become evident that the scientific background of some of these actives is quite complex. This knowledge, however, is fundamental to dermatologists who as "skin experts" are qualified best to provide guidance and advice to the general population on how to prevent premature skin aging.

9.2 Primary Photoprotection: UV Filters

UV filters containing sunscreens have been commercially available for more than 60 years [9]. The safety and efficacy of these products are mainly defined by the type and concentration of the filters employed and how the product has been formulated. UV filters can be divided into physical and chemical filters and most modern sunscreens contain a mixture of both. The general requirements that modern sunscreens should fulfill are the capacity to protect against UVB and UVA radiation, photostability and water resistance, and they should minimize the use of chemical filters and maximize the use of physical filters.

9.2.1 Physical UV Filters

The most frequently used physical UV filters are inorganic micropigments with particle di-

ameters in the range 10 to 100 nm, such as zinc oxide and titanium dioxide [9]. These micropigments are capable of reflecting a broad spectrum of UV rays in the UVA and UVB regions. The current state of research is consistant with the assumption that upon topical application micropigments do not penetrate into viable skin layers and thus have a low potential to exert toxic effects. They are therefore increasingly used in combination with chemical filters to achieve higher sun protection factors and a broader spectrum of protection and to reduce the need for chemical filters.

A major disadvantage of micropigments is that they also reflect visible light, creating the so-called "whitening" or "ghost" effect, which is cosmetically unacceptable and may mean that a sunscreen product containing a micropigment is rejected by the consumer. Also, from a galenic point of view, formluation of micropigments can be a difficult task because they have a strong tendency to agglomerate, which greatly decreases their efficacy.

For the sake of completeness it should be mentioned that in addition to their use as micropigments, titanium dioxide and zinc oxide are also used at larger (>100 nm) particle sizes. These larger particles provide skin protection through both reflection and diffusion of visible light, and so are used make-ups as well as sunscreen products.

9.2.2 Chemical UV Filters

Chemical UV filters have the capacity to absorb short wavelength UV photons and to transform them into heat by emitting long wavelength photons (infrared radiation). Most chemical filters absorb in a relatively small wavelength range. In general, chemical filters may be divided into molecules that absorb primarily in the UVB region (290–310 nm), such as p-aminobenzoic acid derivatives and Zinkacid esters, and molecules that primarily absorb in the UVA region (320–400 nm), e.g. butyl-methoxydibenzoylmethane. In addition, some filter molecules absorb both UVB and UVA photons, e.g. benzophenone. The two most frequently used chemical filters are ethylhexyl-methoxycinnamate and

Table 9.1. Newly developed organic UV filters (*COLIPA* European Cosmetic Perfumery and Toiletry Association, *INCI* International Nomenclature of Cosmetic Ingredients)

COLIPA number	INCI name	Trade name	Protection	
			UVB	UVA
S 71	Terephthalidene dicamphor sulfonic acid	Mexoryl SX		✓
S 73	Drometrizole trisiloxane	Mexoryl XL	✓	✓
S 74	Benzylidene malonate polysiloxane	Parsol SLX	✓	
S 78	Diethylhexylbutamido triazone	Uvasorb HEB	✓	
S 79	Methylene-bis-benzotriazolyl tetramethylbutylphenol	Tinosorb M	✓	
S 80	Disodium phenyl dibenzimidazole tetrasulfonate	Neoheliopan AP		✓
S 81	Bis-ethylhexyloxyphenol methoxyphenyl triazine	Tinosorb S	✓	✓
	Diethylamino hydroxylbenzoyl hexyl benzoate[a]	Uvinul A Plus		✓

[a] EC approval applied for.

butyl-methoxydibenzoylmethane [9]. However, this combination of filter molecules in the same product, renders the whole filter system photounstable, i.e. exposure to UV radiation causes photochemical reactions that generate reactive oxygen species and subsequent phototoxic and photoallergic skin reactions.

In recent years tremendous progress has been made in developing stategies which allow the photostabilization of UV filters. It is, for example, possible to greatly enhance the photostability of the UVA filter butyl-methoxydibenzoylmethane by combining it with other UV filters, such as octocrylene or 4-methylbenzylidene camphor or with non-UV filter molecules such as diethyl-hexyl-2,6-naphthalate. These efforts have greatly improved the efficacy of photoprotection that can be achieved by chemical UV filters.

In this regard it is also worth mentioning that within the last few years new organic UV filters have been developed and seven of these molecules have recently been approved for use in cosmetic products by the European Commission (Table 9.1).

The US Food and Drug Administration has been much slower in approving new chemical sunscreens. Five out of these seven provide significant protection in the UVA range, a fact that reflects the reaction of the chemical industry to numerous studies from the last 10 years that unambigously show that similar to UVB, UVA radiation can cause damage to human skin in-

cluding skin carcinogenesis and premature skin aging [13].

As one consequence there is a growing need for defining standardized and generally accepted methods for the evaluation of UVA photoprotection. In contrast to UVB photoprotection, it has not yet been possible to achieve consensus at an international level as to the method that should be used for the measurement of UVA photoprotection. As a result, different methods are propagated and used in different countries and by different manufacturers. These include in vivo methods, which are especially promoted in Japan and France, and different types of in vitro methods, which are employed not only in Australia, but also increasingly in several European countries. Despite this regulatory problem, however, it is very likely that the increasing use of the new broadband UV absorbers listed in Table 9.1 will greatly improve the efficacy of modern sunscreens [13].

It is important to emphasize that in addition to the absorption properties of a given UV filter the galenic formulation of a sunscreen product is another important factor that determines the efficacy of the product. In this regard the development of the new filter molecules will certainly lead to greater varieties of the characteristics of feel, spreadability and drying of sunscreen products, since four of these molecules are lipophilic and three are water-soluble. Depending on their combination, some of these moleculs are even capable of exerting photoprotective effects in a synergistic manner, and thus by selecting the appropriate combination it should be possible to reduce the overall concentration of chemical filters in the final product [13].

9.3 Secondary Photoprotection

The group of molecules which are being used as actives is extremely heterogeneous and constantly growing in number. Almost "classical" examples are antioxidants, which in addition to vitamins include polyphenols and, in particular, naturally occurring polyphenols such as flavonoids and prozyanidine. These "established" actives are described in greater detail. In addition, selected examples of more recently developed actives, which have photoprotective activities based on completely different molecular targets and mechanisms, are also discussed.

9.3.1 Antioxidants

Antioxidants can serve as hydrogen donors, i.e. they reduce free radicals to less-reactive hydroxy forms. During this process the antioxidant itself changes and transforms into a passive radical. A passive radical is relatively stable and can neutralize a second free radical. Ultimately the antioxidant is consumed. Antioxidants which are typically used in cosmetic products and sunscreens are vitamins and polyphenols.

9.3.1.1 Vitamins E and C

Vitamin E is a prime example of an antioxidant widely found in sunscreen and cosmetic products. Vitamin E is the generic term for a group of tocopherols, which are lipophilic molecules. Alpha-tocopherol has the greatest importance among the naturally occurring tocopherols. Alpha-tocopherol is a hydrochinonether with an unsaturated phytyl rest. Tocopherol acetate is very often the antioxidant used in sunscreen products. It is usually present at concentrations in the range 0.2–1.5 %. Concentrations of less than 0.5 % probably have no measurable antioxidative effect in human skin.

In addition to vitamin E, vitamin C has become more and more important in sunscreen products. In contrast to vitamin E, vitamin C is water-soluble. The excellent reducing capacity of vitamin C makes its use in cosmetic products difficult. Oxidation of vitamin C leads to its fast degradation even before the vitamin C-containing product can be applied to the skin. This degradation process is induced whenever oxygen is present, e.g. during production or storage of vitamin C-containing products and of course immediately before, during and after their application. This problem has prompted the development of stable vitamin C derivatives for use in sunscreen and cosmetic products. The efficacy of some of these has been proven in in vivo studies, and these suggest that these products are able to protect human skin against photoaging.

9.3.1.2 Botanical Antioxidants

The term polyphenols describes molecules with two adjacent hydroxyl groups on a benzol ring. Natural polyphenols are flavonoids and prozyanidine [1]. They are present in numerous foods, e.g. white and red wine, black or green tea, fruits and vegetables. A flavonoid with well-documented photoprotective characteristics is alpha-glycosylrutin [15]. Topical application of this flavonoid prevents photoprovocation-induced skin lesions in patients with polymorphous light eruption [14]. In vitro studies have demonstrated that supplementation of human dermal fibroblasts with alpha-glycosylrutin protects these cells against UV-induced upregulation of matrix metalloproteinase-1 (MMP-1) and MMP-2 [15].

In addition to single polyphenols, polyphenolic plant extracts have been assessed for their photoprotective potential. Examples include extracts from grape seeds, which primarily contain prozyanidine dimers, from green tea or from gingko biloba. These examples illustrate the increasing use of botanical antioxidants in modern photoprotection. This is a trend that is still ongoing, and some of the most prominent botanical antioxidants are therefore discussed in greater detail.

Tea Polyphenols

Two-thirds of the world's population regularly drink tea from *Camelia sinensis*, thereby making tea, after water, the most popular drink. Tea leaves are processed in several different ways and are commercially available either as black tea (78%), green tea (20%) or Oolong tea (2%). Black tea is primarily popular in Western countries, whereas green tea is mainly consumed in Japan, China, Korea, India and some countries of North Africa and the Middle East. Oolong tea, which is partially fermented, is particularly popular in southeast China.

Green tea polyphenols have been intensively studied in recent years for their photoprotective capacity, and as a consequence are now often found in sunscreen and cosmetic products. The major green tea polyphenols are:

– (–)-Epigallocatechin-3-gallate (EGCCG)
– (–)-Epigallocatechin (EGC)
– (–)-Epicatechin-gallate (ECG)
– (–)-Epicatechin
– (+)-Gallocatechin
– (+)-Catechin

In addition, black tea contains thearubigin and theaflavin, which are generated during enzymatic oxidation.

All of these polyphenols are potent antioxidants. For green tea polyphenols, both topical application and oral uptake have been demonstrated to be photoprotective. Accordingly, green tea polyphenols effectively prevent UVB-induced DNA damage, UVB-induced skin cancer, and UVB-induced immunosuppressive effects [1]. In addition, there is good evidence that green tea polyphenols provide protection against UV-induced premature skin aging.

In many aspects, green tea polyphenols resemble retinoids. For example, it has been shown that theaflavin and EGCG inhibit the UVB-induced activation of transciption factor AP-1 in vitro in primary human keratinocytes and in vivo in transgenic mice [2]. AP-1 activation is of central importance for UVB-induced MMP-1 expression and thereby for collagen degradation, indicating that these molecules may have anti-photoaging properties [1].

Topical application of green tea polyphenols or EGCG alone prevents a number of UVB-induced actinic changes in human skin, including erythema (i.e. sunburn), the development of an inflammatory infiltrate, the generation of sunburn cells, an increase in prostaglandin synthesis, and the induction of myeloperoxidase activity [5]. Other protective effects include the reduced production of reactive oxygen species and lipid peroxidation products, a reduced depletion of epidermal Langerhans cells and of endogenous anitioxidant systems. Taken together these studies indicate that green tea polyphenols have a strong photoprotective potential and should be effective in preventing skin photoaging [1]. Accordingly, very recent studies have demonstrated that oral administration of green tea polyphenols to SKH-1 hairless mice effectively prevents UVB-induced (chronic, repetitive exposure) protein oxidation, a hallmark of photoaged skin, and UVB-induced expression of matrix-degrading MMP-2 (gelatinase A), MMP-7, and MMP-9 (gelatinase B), that degrades collagen fragments of type I and III generated by collagenase, and MMP-3 (stromelysin), that degrades type IV collagen of the basement membrane [18].

Curcumin

Curcumin, which is best known as a spice and colour additive in food from India, is also a potent antioxidant with antiinflammatory and antiproliferative properties. Similar to green tea polyphenols, curcumin has the capacity to prevent the activation of the transcription factors AP-1 and NFκB, which are critically involved in a variety of UV-induced biological processes in the skin. There are also numerous reports that indicate that curcumin can be used in the chemoprevention of UV-induced tumors [1].

A major disadvanatge of curcumin, however, is its yellow colour, which discourages its use in cosmetic preparations.

Silymarin

Silymarin is a polyphenolic flavonoid. It is isolated from *Silyum marianum* and represents a mixture of different flavolignanes including silybin, silydinanin, silycristin and isosilybin. A number of studies have shown that silymarin, similar to green tea polyphenols and curcumin, is able to protect against UV-induced skin cancer in mice [1]. In particular, topical application of silymarin reduces UV-induced edema, sunburn cells, depletion of antioxidant systems such as catalase and UV-induced cyclooxygenase induction. Analysis of the underlying molecular mechanisms has demonstrated that NFκB inhibition is a major mechanism responsible for the photoprotective effects induced by silymarin. Based on these properties, it is likely that silymarin also provides protection against UV-induced premature skin aging.

Apigenin
Apigenin is a natural flavonoid present in a number of fruits and vegetables including apples, beans, broccoli, cherries, grapes, onions, parsely and tomatoes. Tea and wine also contain apigenin. In comparison to other flavonoids, apigenin is less toxic and has a low mutagenic potential. Topical application of apigenin reduces UVB-induced skin cancer in mice [1].

Resveratrol
Resveratrol is a polyphenolic phytoalexin, which is present in grapefruit, nuts, fruits and red wine, and it exhibits potent antiinflammatory and antiproliferative capacities. Accordingly, application of resveratrol to the skin of hairless SKH-1 mice effectively prevented the UVB-induced increase in skin thickness and the development of skin edema [1]. Immunohistochemical studies have revealed that this inhibitory effect is associated with a reduction in UVB-induced hydroperoxide formation and the subsequent formation of leukocyte infiltrates in the UVB-irradiated skin area. Topical application of resveratrol also inhibits the UVB-induced formation of lipid peroxides. It is likely that this substance has the capacity to provide protection against UV-induced premature skin aging as well.

9.3.1.3 Carotenoids

Carotenoids such as β-carotene or lycopene are highly effective antioxidants and neutralize singlet oxygen and peroxyradicals which are frequently formed during photooxidative processes. A number of recent studies in humans have shown that oral uptake of carotenoids has photoprotective effects [16]. For patients with erythropoietic protoporphyria oral administration of β-carotene is the measure of choice to protect against UV-induced skin damage. In normal healthy volunteers a significant reduction in solar simulator-induced erythema formation has been observed after 12 weeks oral administration of β-carotene alone and after 8 weeks oral administration of β-carotene together with tocopherol. Similar effects have also been achieved in volunteers kept on a lycopene-rich diet [16]. Oral uptake of tomato paste in quantities representing 16 mg lycopene per day over a total period of 10 weeks caused a significant increase in lycopene serum levels and of the total carotene content in skin. This increase was functionally relevant because 10 weeks after initiation of this diet UV-induced erythema formation (sunburn) was significantly reduced, in comparison to a placebo-treated group [16].

These studies suggest that oral uptake of selected (micro)nutrients can provide photoprotection of human skin. However, we should maintain a healthy scepticism, since many well-designed studies have also failed to find that oral antioxidants protect against DNA damage, cancer and other endpoints [11]. In this regard it is, however, important to emphasize that with these molecules photoprotection can only be achieved if an optimal dose range is reached in the skin. In fact, in vitro studies clearly indicate that excessive carotenoid concentrations may even have opposite, i.e. prooxidative, effects.

The relevance of carotenoid-induced photoprotection for prevention of photoaging is suggested by very recent studies in which β-carotene, lycopene, lutein, zeaxanthin and cryptoxanthin have been analyzed for their capacity to prevent premature skin aging in a well-established, standardized human in vitro model in which repetitive thrice-daily exposure of primary human dermal fibroblast to very low subtoxic doses of UVA radiation led to the generation of mutations in mitochondrial DNA. In this system, supplementation of fibroblasts with these antioxidants provided significant protection against UVA radiation-induced mitochondrial DNA mutagenesis (Krutmann et al., unpublished data; [4]). It was also observed that

9

the photoprotective effectiveness of these molecules depends on the type of carotenoid used and that for each molecule a critical dose range had to be defined at which photoprotection can be achieved without promutagenic effects from the carotenoids alone.

9.3.1.4 *Phyllantus emblica*

The fact that antioxidants may also be prooxidative has prompted the development of novel antioxidants which do not exert prooxidative effects even when used at high concentrations. In this regard a prototpye is a standardized extract from the tree *Phyllantus emblica* which is commercially available as Emblica [3]. *Phyllantus emblica* is one of the most important ayurvedic herbs from India and has been used in traditional Indian medicine for thousands of years. More recent studies have assessed whether Emblica extracts can be used for skin-care and skin-protection purposes. In aggregate, these studies indicate that this extract has very potent antioxidative properties and does not exert prooxidative effects, even when it is employed at very high concentrations. In addition to its antioxidant effects, this extract has been found to effectively chelate Fe^{3+} and Cu^{2+} and to prevent the formation of hydroxyradicals. Another interesting property of Emblica extract seems to be that, in contrast to other antioxidants, it is not transformed upon exposure to oxidative stress from an active into an inactive form. Instead, evidence has been obtained that the extract contains a cascade of different, interacting components with antioxidative properties which seems to be responsible for its relatively long-lasting antioxidant activity [3].

9.3.1.5 L-Ergothioneine

L-Ergothioneine is a naturally occurring, non-essential amino acid with strong antioxidant activity due to is thione group [6]. It is synthesized exclusively by fungi and mycobacteria, and it is taken up by humans in plant foods and avidly conserved. It is found in millimolar quantities in blood (especially erythrocytes), seminal fluid

and the eye lens. It is a superoxide and singlet-oxygen scavenger, and it reduces the expression of UV-induced inflammatory cytokines and metalloproteases associated with photoaging [12]. L-Ergothioneine is finding increased popularity in cosmetic products, particularly where it can recycle spent vitamin C.

9.3.2 Osmolytes

Exposure of cells to osmotic stress, e.g. hyperosmotic cell shrinkage or hypoosmotic cell swelling, leads to the uptake or efflux of compatible organic solutes or osmolytes in order to keep the cell volume constant. Osmolytes found in human cells include betaine, taurine and myoinositol. During recent years numerous studies have demonstrated that osmolytes not only play a role in regulation of cell volume homeostasis but are in fact an integral part of a cell's defense system against exogenous noxae. The molecular basis of this so-called "osmolyte strategy" is a system of specific osmolyte transport systems, e.g. the betaine gamma-amino-N-butyric acid transporter-1 (BGT-1), the sodium-dependent myoinositol transporter (SMIT) and the taurine transporter (TAUT). The expression and function of these transporters has initially been characterized in liver, kidney and neural cells and these studies have demonstrated that each of these tissues uses a specific osmolyte strategy. Very recently Warskulat et al. have reported that epidermal keratinocytes are equipped with an osmolyte strategy as well [20].

Accordingly, primary normal human epidermal keratinocytes have been found to express mRNA for the osmolyte transporters BGT-1, SMIT and TAUT. Expression of these transporters is of functional relevance because hyperosmotic stimulation of keratinocytes led to an uptake of the respective osmolytes, whereas hypoosmotic stress had the opposite effect. It has been known for some time that oxidative stress can alter the cellular hydration state. Since UV radiation is a major source of oxidative stress, the authors next assessed whether UV irradiation affects osmolyte uptake or efflux by human keratinocytes. They observed that both UVB and UVA radiation induce the uptake of betaine, myoinositol and, in particular, taurine, whereas osmolyte efflux is not affected. These results indicate that osmolyte uptake is part of the UV radiation-induced stress response in human skin. Additional studies have revealed that osmolyte, and in particular taurine, uptake is part of the natural defense system of skin cells against UV radiation-induced detrimental effects. Pretreatment of primary human epidermal keratinocytes in vitro with taurine provides almost complete protection against UVB- and UVA-induced expression of the proinflammatory molecule ICAM-1 and the cytokines TNF-alpha and IL-10, which are known to play a pivotal role in mediating UVB-induced immunosuppressive effects. Also, TAUT knockout mice, which have >90 % reduced skin taurine levels, are significantly more sensitive to UVB radiation-induced immunosuppression in vivo, as compared with wild-type mice (Krutmann et al., unpublished data; [20]).

These and additional in vitro studies have prompted the increasing use of osmolytes as actives in cosmetic and sunscreen products. In addition to taurine and betaine, which are both osmolytes that are physiologically used by the skin, an increasing number of products contain ectoin. Ectoin is not produced by human cells, but is synthesized by halophilic bacteria such as *Ectothiospira halochloris* [2]. These bacteria are present in salt-containing alkaline lakes, and are thus subject to strong osmolarity variations in their environment as well as high doses of UV radiation. In vivo studies in humans have shown that topical application of ectoin protects against UVB-induced Langerhans cell depletion and sunburn cell formation. Preincubation of human epidermal keratinocytes with ectoin completely prevents UVA radiation-induced signal transduction and gene expression through a mechanism which involves the stabilization of signal-transducing lipid rafts in the keratinocyte cell membrane [2]. These studies clearly indicate that the photoprotective activity of osmolytes is not due to an antioxidant effect, but is based on completely different molecular mechanisms. Osmolytes may also be capable of preventing UV radiation-induced premature skin aging, because pretreatment of dermal fibroblasts with ectoin or taurine significantly

reduces UVA radiation-induced mitochondrial DNA mutagenesis as well as MMP-1 expression, which are models for photoaging (Krutmann et al., unpublished data). It is anticipated that osmolytes will become of increasing importance as a novel group of actives for photoprotective/antiaging strategies. In addition, further analysis of osmolyte strategies that are physiologically employed by skin cells will identify novel strategies and targets for the development of new actives.

9.3.3　DNA Repair Enzymes

Preventing DNA damage by blocking or filtering UV photons and intercepting oxygen radicals is clearly central to photoprotection. An additional strategy is to increase repair of DNA damage after it occurs but before it can cause biological consequences. Over the past 40 years in the field of DNA repair many enzymes that recognize and initiate removal of DNA damage, by nucleotide excision repair, base excision repair or direct reversal, have been identified. The use of some of these enzyme activities for photoprotection became practical with the development of liposomes specifically engineered for delivery into skin [23]

The small protein T4 endonuclease V from bacteriophage recognizes the major form of DNA damage produced by UVB, which is the cyclobutane pyrimidine dimer (CPD). A similar enzyme is found in extracts of *Micrococcus luteus*. Liposomal delivery of T4 endonuclease V to UV exposed human skin increases repair from 10% of CPD to 18% over 6 hours, but dramatically reduces or eliminates the release of cytokines such as IL-10 and TNFα [21]. In a randomized clinical study of the effects of daily use of this liposomal T4 endonuclease V in patients with xeroderma pigmentosum, the rate of premalignant actinic keratosis and basal cell carcinoma was reduced by 68% and 30%, respectively, compared to the placebo control [24].

The enzyme photolyase, found ubiquitously in plants, reptiles, amphibians and marsupials, directly reverses CPD by absorbing visible light and using the energy to split the cyclobutane ring. Liposomal delivery of photolyase to

Table 9.2. 8-Oxo-guanine in normal human epidermal keratinocytes after treatment with liposomal OGG1

Treatment	Liposome	Repair time (h)	$8oG/10^6\text{-base}^a$
Untreated	None	0	0.0
H_2O_2	None	0	5.0
H_2O_2	None	2	1.9
H_2O_2	None	6	2.0
H_2O_2	OGG1	2	0.0
H_2O_2	OGG1	6	0.0
H_2O_2	Empty	2	1.7
H_2O_2	Empty	6	1.3

[a] 8-Oxo-guanine measured by the endonuclease-sensitive-site assay, corrected for background.

UV-irradiated dendritic cells of mouse skin restores their antigen-presenting activity [19], and treatment of UV-irradiated human cells reduces apoptosis [8]. Treatment of UV-irradiated human skin with liposomal photolyase reduces the level of CPD [17] and the degree of suppression of the contact hypersensitivity response [15].

Oxygen radicals most often oxidize the guanine base of DNA to form 8-oxo-guanine (8oG), which is a mutagenic lesion. A widespread enzyme that initiates base excision repair of this damage is oxoguanine glycosylase 1 (OGG1). Liposomal delivery of the *Arabidopsis* OGG1 enzyme (30 ng/ml) to human epidermal keratinocytes dramatically increases their repair of 8oG induced by hydrogen peroxide (H_2O_2/ $FeSO_4$/$CuSO_4$; Table 9.2). Repair of 8oG occurs over 2–6 hours in untreated cells, or in cells treated with empty liposomes not containing any enzyme. Treatment with OGG1 liposomes eliminates 8oG within 2 hours of treatment.

The delivery of biologically active repair enzymes into skin using liposomes provides an important modality to provide photoprotection even after the initial insult has occurred.

9.4 Concluding Remarks

The aging of the populations in the industrial world means that we are living much longer with the consequences of sun exposure. The first approaches of screening or filtering the solar rays are chemically and physically effective but ran into resistance from human behaviour.

A better understanding of the biochemistry led to the use of a variety of antioxidants which quench or intercept oxygen radicals and ameliorate some of the effects of solar UV. We now have available DNA repair enzymes to actually reverse or repair damage after it happens but before it can have its effects. We should be careful to use these tools properly, so that we do not facilitate greater overall exposure to some or all of the solar spectrum in the mistaken belief that protection is perfect. But in societies where the burden of medical care is becoming overwhelming, we should remember that prevention is preferable, both economically and socially, to treatment.

References

1. Afaq F, Mukhtar H (2002) Photochemoprevention by botanical antioxidants. Skin Pharmacol Appl Skin Physiol 15:297–306
2. Buenger J, Driller H (2004) Ectoin: an effective natural substance to prevent UVA-induced premature photoaging. Skin Pharmacol Physiol 17:232–237
3. Chaudari RK (2002) Emblica cascading antioxidant: a novel natural skin care ingredient. Skin Pharmacol Appl Skin Physiol 15:374–380
4. Eicker J, Kürten V, Wild S, Riss G, Goralczyk R, Krutmann J, Berneburg M (2003) Betacarotene supplementation protects from photoaging-associated mitochondrial DNA mutation. Photochem Photobiol Sci 2:655–659
5. Elmets CA, Singh D, Tubesing K, Matsui M, Kotiyar SK, Mukhtar H (2001) Cutaneous photoprotection from ultraviolet injury by green tea polyphenols. J Am Acad Dermatol 44:425–432

9

6. Hartman P (1990). Ergothioneine as antioxidant. Methods Enzymol 186:310–318

7. Krutmann J (2001) New developments in photoprotection of human skin. Skin Pharmacol Appl Skin Physiol 14:401–407

8. Kulms D, Pöppelmann B, Yarosh D, Luger TA, Krutmann J, Schwarz T (1999) Nuclear and cell membrane effects contribute independently to the induction of apoptosis in human cells exposed to UVB radiation. Proc Natl Acad Sci U S A 96:7974–7979

9. Lowe NJ, Shauth NA, Patahk MA (1997) Sunscreens – development, evaluation and regulatory aspects. Marcel Dekker, New York

10. Mahroos M, Yaar M, Phillips T, Bhawan J, Gilchrest B (2002) Effect of sunscreen application on UV-induced thymine dimers. Arch Dermatol 138:1480–1485

11. Møller P, Loft S (2004) Interventions with antioxidants and nutrients in relation to oxidative DNA damage and repair. Mutat Res 551:79–89

12. Obayashi K, Kurihara K, Okano Y, Masaki H, Yarosh DB (2005) L-Ergothioneine scavenges superoxide and singlet oxygen and suppresses TNF-α and MMP-1 expression in UV-irradiated human dermal fibroblasts. J Cosmet Sci 56:17–27

13. Osterwalder U, Luther H, Herzog B (2001) Sun protection beyond the sun protection factor – new efficient and photostable UVA filters. SÖFW J 127:45–54

14. Rippke F, Wendt G, Bohnsack K, Dörschner A, Stäb F, Hölzle E, Moll I (2001) Results of photoprovocation and field studies on the efficacy of a novel topically applied antioxidant in polymorphous light eruption. J Dermatol Treat 12:3–8

15. Stäb F, Wolber R, Mundt C, Blatt T, Will K, Keyhani R, Rippke F, Max H, Schönrock U, Wenck H, Moll I, Hölzle E, Wittern K-P (2001) Alpha-glycosylrutin – an innovative antioxidant in skin protection. SÖFW J 127:2–8

16. Stahl W, Sies H (2002) Carotenoids and protection against solar UV radiation. Skin Pharmacol Appl Skin Physiol 15:291–296

17. Stege H, Roza L, Vink A, Grewe M, Ruzicka T, Grether-Beck S, Krutmann J (2000) Enzyme plus light therapy to repair DNA damage in ultraviolet-B-irradiated human skin. Proc Natl Acad Sci U S A 97:1790–1795

18. Vayalli PK, Mittal A, Hara Y, Elmets CA, Katiyar SK (2004) Green tea polyphenols prevent ultraviolet light-induced oxidative damage and matrix metalloproteinase expression in mouse skin. J Invest Dermatol 12:1480–1487

19. Vink A, Moodycliffe A, Shreedhar V, Ullrich S, Roza L, Yarosh D, Kripke M (1997) The inhibition of antigen-presenting activity of dendritic cells resulting from UV irradiation of murine skin is restored by in vitro repair of cyclobutane pyrimidine dimers. Proc Natl Acad Sci U S A 94:5255–5260

20. Warskulat U, Reinen A, Grether-Beck S, Krutmann J, Häussinger D (2004) The osmolyte strategy of human keratinocytes in maintaining cell homeostasis. J Invest Dermatol 123:516–521

21. Wolf P, Maier H, Müllegger R, Chadwick C, Hofmann-Wellenhof R, Soyer HP, Hofer A, Smolle J, Horn M, Cerroni L, Yarosh D, Klein J, Bucana C, Dunner K, Potten C, Hönigsmann H, Kerl H, Kripke M (2000) Topical treatment with liposomes containing T4 endonuclease V protects human skin in vivo from ultraviolet-induced upregulation of interleukin-10 and tumor necrosis factor-α. J Invest Dermatol 114:149–156

22. World Health Organization (2001) Sunscreens. IARC Handbooks of Cancer Prevention, vol 5. International Agency for Research on Cancer, Lyon, France

23. Yarosh D, Bucana C, Cox P, Alas L, Kibitel J, Kripke M (1994) Localization of liposomes containing a DNA repair enzyme in murine skin. J Invest Dermatol 103:461–468

24. Yarosh D, Klein J, O'Connor A, Hawk J, Rafal E, Wolf P (2001) Effect of topically applied T4 endonuclease V in liposomes on skin cancer in xeroderma pigmentosum: a randomized study. Lancet 357:926–929

Chapter 10

Vitamins and Polyphenols in Systemic Photoprotection

10

Wilhelm Stahl, Hasan Mukhtar, Farrukh Afaq, Helmut Sies

Contents

10.1 Introduction

The human organism depends on an adequate energy supply provided by major dietary components, protein, carbohydrates and lipids. However, minor constituents such as vitamins, minerals and specific fatty acids are required in a healthy diet as well. Secondary plant compounds are ingested with food and enter the systemic circulation. These are not essential in the strict sense of a vitamin, but some of these compounds exhibit distinct biological activities. Among them are terpenoids and polyphenols such as carotenoids, tocopherols, and flavonoids [1–3] which are known to be efficient antioxidant micronutrients.

As the exterior barrier of the body, the skin is in direct contact with the environment. This organ is exposed to oxygen and light, conditions under which reactive oxygen species (ROS) are generated. Photooxidative stress is involved in processes of photoaging and photocarcinogenesis, and plays a major role in the pathogenesis of photodermatoses [4].

As any other tissue, skin depends on an optimal supply of nutritive compounds. Skin benefits from dietary antioxidants capable of scavenging reactive intermediates generated under the condition of photooxidative stress [5–7]. Micronutrients may also act as UV absorbers, or modulate signaling pathways elicited upon UV exposure [8–10].

In plants, minor constituents play an important role in protection against excess light. Besides acting as accessory pigments, carotenoids are associated with photoprotection [11], being involved in the dissipation of excess light energy through the xanthophyll cycle, quenching excited triplet state molecules and singlet oxygen. Based on their structural features which determine their physicochemical properties, carotenoids, flavonoids, and vitamins E and C are also suitable compounds for photoprotection in humans [6].

10.2 Systemic Photoprotection

Topically applied sunscreens protect against UV-induced skin lesions and prevent sunburn. Sun protection factors (SPF) of 20 or more are achieved if sunscreens are properly used. In contrast to topical protection, the concept of endogenous protection implies systemic delivery of the protective agent and its transport to the target site. The concept is attractive since it overcomes problems associated with topical sun protection, but it also shows several limitations with regard to efficacy, toxicity, and availability. Systemic photoprotection is convenient and may provide a desirable basic UV shield for the entire body surface. Endogenous skin protection contributes to the protection of sensitive dermal target sites beyond those reached by sunscreens. Although endogenous protection in terms of SPF may be low compared to sunscreens, the potential cumulative effect is noteworthy because lifelong inadvertent exposure to sunlight is substantial [12]. The average annual dose of erythemal UV in the US is about 25,000 J/m^2 per year, 22,000 J/m^2 for females and 28,000 J/m^2 for males [13, 14]. An additional dose of about 7,800 J/m^2 can be received during a conventional vacation. Thus, most exposure to UV light occurs under everyday circumstances, when no topical protection by sunscreens is used.

It is important to note that endogenous photoprotection is complementary to topical photoprotection, and these two forms of prevention clearly should not be considered mutually exclusive. Promising candidates for endogenous photoprotection are found among the antioxidant micronutrients including carotenoids, tocopherols, flavonoids, other polyphenols, and vitamin C [6].

The structural requirements for a suitable photoprotective compound depend on the supposed underlying mechanism of action [15]. UV-absorbing molecules with a high extinction coefficient and a broad absorption maximum in the UV range may be used to increase the barrier against UV light. Antioxidants act as scavengers of ROS generated in primary or secondary reactions following UV irradiation. Repair of UV-induced damage may be supported by molecules which are capable of inducing suitable repair systems; antiinflammatory agents can be used to suppress cellular responses.

The concept of endogenous photoprotection implies that the compound or its biologically active metabolite is available in sufficient amounts at the target site [16]. Thus, pharmacokinetic parameters including absorption, distribution, metabolism, and excretion may affect the level of the compound in the skin and its bioactivity [17, 18].

For endogenous use, compounds need to be safe in the recommended dose range and over a longer period of time. Safety concerns have been raised for β-carotene, one of the most frequently applied endogenous sun protectants. There is evidence from two intervention trials that the risk of lung cancer is increased in high-risk populations, such as smokers and asbestos workers, following high-dose supplementation with β-carotene [19, 20]. The unexpected results from these intervention trials have been broadly discussed, and several hypotheses have been raised to explain their outcome.

10.3 Vitamins E and C

The term vitamin E is a generic description for all tocols and tocotrienol derivatives which exhibit the biological activity of α-tocopherol [17]. This group of compounds is highly lipophilic, operative in membranes and lipoproteins. Their most important antioxidant function appears to be the inhibition of lipid peroxidation, scavenging lipid peroxyl radicals to yield lipid hydroperoxides and a tocopheroxyl radical [21–23].

Vitamin C (ascorbate) is a water-soluble compound, and the most important reaction requiring ascorbate as a cofactor is the hydroxylation of proline residues in collagen. Vitamin C is essential for the maintenance of normal connective tissue as well as for wound healing [24]. Ascorbate is considered one of the most powerful least toxic natural antioxidants and is present in high concentrations in many tissues. Upon interaction with ROS it is oxidized to dehydroascorbate via the intermediate ascorbyl free radical. As a scavenger of ROS, ascorbate has been shown to be effective against superoxide radical anion, hydrogen peroxide, the hydroxyl radical, and singlet oxygen [25].

In a study investigating the cooperative activity of the two compounds, vitamin E and vitamin C, against UV-induced erythema [26], four treatment groups were followed over a period of 50 days. The subjects received (R,R,R)-α-tocopherol and ascorbate supplementation as single components at dose levels of 2 g/day and 3 g/day, respectively. Combination treatment comprised 2 g/day of α-tocopherol plus 3 g/day of ascorbate; controls remained without treatment. After treatment with the combination of the two vitamins the "sunburn threshold" or minimal erythemal dose (MED) was significantly increased from about 100 mJ/cm^2 before to about 180 mJ/cm^2 after supplementation. Treatment with the single compounds provided only moderate protection, which was not statistically significant.

Short-term intervention with high doses of vitamin E and vitamin C provides some protection. When 1,000 IU vitamin E was ingested together with 2 g ascorbic acid per day over a period of 8 days, a minor increase in MED was seen [27].

(R,R,R)-α-Tocopherol supplementation in combination with carotenoids has been used as an oral sun protectant. Supplementation with vitamin E and β-carotene tended to be superior to β-carotene treatment alone, but the difference was statistically not significant [28]. When vitamin E (α-tocopheryl acetate) was administered alone at a dose level of 400 IU/day over a period of 6 months, no significant protection was achieved [29].

Little is known about the bioavailability, metabolism and distribution of tocopherols in human skin or parameters determining the distribution in different skin layers. However, it has been shown that individual MED is not correlated with the epidermal content of tocopherols [30]. It is noteworthy that in addition to transport via epidermal blood vessels, α- and γ-tocopherol are continuously secreted with

human sebum [31]. The stratum corneum is the first line of defense against exogenous noxae including oxidants such as ozone, and vitamin E is an important constituent of the antioxidant network in this skin layer [32, 33].

In a study with 12 volunteers, the subjects received vitamin C supplementation at 500 mg/day over a period 8 weeks, and the erythemal response following UV exposure was followed [34]. Supplementation with only vitamin C had no effect on MED.

Human and animal demonstrate that vitamin E and, to a lesser extent, vitamin C, also provide UV protection when the compounds are applied topically [35, 36]. Several studies have shown that vitamins E and C may prevent phototoxic damage including UV-dependent erythema, formation of sunburn cells, lipid peroxidation and DNA damage [37–39]. Usually a combination of the vitamins is more efficient than the single compounds as ingredients of a topically applied sunscreen [35, 40].

10.4 Carotenoids

Carotenoids are among the most widespread natural colorants, with β-carotene as the most prominent [41, 42]. β-Carotene supplements are frequently used as so-called oral sun protectants, but studies proving a protective effect of oral treatment with β-carotene against skin responses to sun exposure are scarce, and conflicting results have been reported. Importantly, at present, positive studies suggesting efficacy have examined only erythema, and no data support a reduction in UV-induced damage to DNA or other skin constituents.

In human skin, carotenoid levels have been determined to be in the range 0.2 to 0.6 nmol/g wet tissue, not including subcutaneous fat which generally contains much higher levels [43]. There are striking differences in the levels of single carotenoids and in the distribution of carotenoids in different skin areas [44]. After supplementation with 24 mg β-carotene (from the alga *Dunaliella salina*) over a period of 12 weeks, carotenoid skin levels increased significantly [44]. Increasing levels of carotenoid were measured in all areas of the skin, but the most pronounced increases were in the skin of the forehead, back, and palm of the hand. When treatment with β-carotene was discontinued after 12 weeks, carotenoid levels decreased in all skin areas.

In an intervention study with the same preparation and dose of β-carotene it was shown that protection against UV-induced erythema can be achieved [28]. As expected, carotenoid levels in the skin and serum rose upon supplementation. The protective effects of treatment were evaluated by measuring the intensity of the erythema following UV exposure before and during treatment. Compared to baseline, erythema intensity was significantly diminished from week 8 onwards. The effect was even more pronounced after 12 weeks of treatment, in particular when the subjects received supplementation with a combination of carotenoids and vitamin E.

The efficacy of β-carotene in systemic photoprotection depends on the duration of treatment and on the dose. In studies documenting protection against UV-induced erythema, supplementation with carotenoids lasted for more than 8 weeks, and the dose was at least a total of 12 mg carotenoids per day [28, 45–48]. In studies showing no protective effects the treatment period was shorter [49–51].

Supplementation with β-carotene at high doses is controversial because of safety concerns [52]. In two intervention trials with individuals at a high risk of lung cancer, a higher cumulative index for lung cancer and certainly no protective effect was observed in the groups that received β-carotene [19, 53]. No elevated risk was found in three other intervention trials with comparable doses of β-carotene.

A high dose of β-carotene can be substituted by a mixture of different carotenoids that retains the sun-protective effects [46]. In this study the participants received either β-carotene at a dose of 24 mg/day or a carotenoid mixture providing 24 mg of total carotenoids per day, for a period of 12 weeks. The carotenoid mixture consisted of β-carotene, lycopene, and lutein, with 8 mg of each compound. In the β-carotene group, only serum levels of β-carotene increased substantially whereas an increase of all three carotenoids, β-carotene, lycopene, and lutein, was found in the group which ingested the carotenoid mixture. Increases in total carotenoids of the skin were comparable in both groups. Erythema intensity in response to UV-irradiation was diminished in both groups that received carotenoids and was significantly lower than baseline after 12 weeks of supplementation. Thus it was concluded that long-term supplementation with a mixture of carotenoids supplying equal amounts of β-carotene, lutein and lycopene ameliorates UV-induced erythema in humans [46]. The protection provided by the mixture of carotenoids is comparable to that observed after treatment with 24 mg of β-carotene alone.

Fruits and vegetables provide most of the carotenoids in the human diet with a mean daily intake of about 5 mg total carotenoids per day, as measured in Germany [54]. In contrast to many other carotenoids, lycopene is present only in a few food items. Tomatoes and tomato products are the major lycopene sources in a Western diet, providing more than 90% of total lycopene intake.

In order to investigate whether a diet rich in carotenoids is useful for photoprotection, tomato paste was selected as a dietary source containing high amounts of lycopene [55]. To improve absorption, olive oil was coingested with the paste (20 g/day), providing 16 mg lycopene per day. Dietary intervention was performed over a period of 10 weeks, after which increased lycopene levels in serum and skin were found. After 10 weeks of intervention erythema formation was significantly lower in the group consuming the tomato paste compared to control [56], but no significant protection was found during week 4 of treatment.

The results show that protection against UV-induced erythema can be achieved by modulation of the diet. The first approaches to developing functional foods with carotenoids as sun-protective agents have shown promising results [57]. Future studies to determine whether oral carotenoids can reduce UV-induced tissue damage will be of interest.

10.5 Polyphenols

10.5.1 Green Tea Polyphenols

Green tea contains four major types of polyphenols: (–)-epicatechin (EC), (–)-epicatechin gallate (ECG), (–)-epigallocatechin (EGC) and (–)-epigallocatechin-3-gallate (EGCG). All of these green tea polyphenols (GTP) act as potent antioxidants and can scavenge ROS, such as lipid free radicals, superoxide radicals, hydroxyl radicals, hydrogen peroxide and singlet oxygen. Treatment of cultured cells with EGCG, directly inhibits the baseline expression of matrix metalloproteinases (MMPs) such as MMP-2, MMP-9 and neutrophil elastase even in the absence of UV exposure [58]. Chronic exposure of mouse skin to UVB has been shown to induce the expression of MMP-2, -3, -7, and -9, which are involved in the degradation of types I and III collagen fragments generated by collagenases [59], and type IV collagen of the basement membrane [60]. Oral administration of GTP to mice as a sole source of liquid markedly inhibits UV-induced expression of MMPs in the skin, suggesting that GTP has a potential anti-photoaging effect [59]. Treatment with EGCG diminishes UVA-induced skin damage (roughness and sagginess) and protects against the loss of dermal collagen from the skin in hairless mice, and also blocks the UV-induced increase in collagen secretion and collagenase mRNA levels in cultured human epidermal fibroblasts and the promoter-binding activities of AP-1 and NF-κB [61]. In rats, feeding a green tea extract markedly inhibits the age-associated increase in the fluorescence in the aortic collagen [62].

Oxidative damage induced by UVB may also cause modifications of proteins, e.g., collagen crosslinking and formation of carbonyl derivatives. In a study with mice it was found that collagen crosslinking could be diminished by green tea extract. EGCG topically applied to mice, before exposure to a single dose of UVB irradiation, inhibits hydrogen peroxide and nitric oxide production in both the dermis and epidermis [63]. In mice, topical application of EGCG to the skin prior to UV irradiation leads to an inhibition of contact hypersensitivity responses,

a decreased number of infiltrating macrophages and neutrophils (CD11b$^+$ cells), downregulation of UV-induced production of interleukin-10 (IL-10) and an increased production of IL-12 in the skin and draining lymph nodes [64]. EGCG has also been found to balance the alterations in IL-10/IL-12 which may be mediated by the antigen-presenting cells (APCs) in the skin and draining lymph nodes or by blocking the infiltration of IL-10 secreting CD11b$^+$ macrophages into the irradiated site. EGCG also inhibits the migration, depletion and death of APCs when detected as class II MHC$^+$Ia$^+$ cells, and significantly decreases dermal and epidermal H$_2$O$_2$ and NO production [64].

Pretreatment of normal human epidermal keratinocytes with EGCG inhibits UVB-induced oxidative stress-mediated phosphorylation of MAPK, phosphorylation and degradation of IκBα and activation of IKKα and NF-κB [65, 66]. In the human keratinocyte cell line HaCaT, EGCG inhibits UVB-mediated activation of AP-1 activity [67]. EGCG also inhibits UVB-induced expression of c-fos, a major component of AP-1 [68]. Topical application of green tea polyphenols to SKH-1 hairless mice prior to multiple UVB exposure downregulates UVB-induced phosphorylation of MAPK and activation of NF-κB [69]. In a study including 118 patients with atopic dermatitis, the consumption of three cups of oolong tea (a combination of green and red tea) for 6 months decreased the severity of the disease [70].

10.5.2 Silymarin

Silymarin, a polyphenolic flavonoid isolated from the milk thistle plant (*Silybum marianum* L. Gaertn), is a mixture of several flavonolignans, which include silybin, silybinin, silydianin, silychristin and isosilybin [71]. The antioxidant effects of silymarin have been established in the mouse model, and silybin is the main constituent responsible for these effects [71]. Topical application of silymarin protects against photocarcinogenesis in mice: silymarin applied to SKH-1 hairless mice prior to UVB irradiation significantly reduced tumor incidence, tumor multiplicity, and average tumor volume

10

[72]. Further, in short-term experiments, topical application of silymarin was found to result in significant inhibition against UVB-induced skin edema, formation of sunburn and apoptotic cells, depletion of catalase activity, and induction of COX and ODC activities and ODC mRNA expression [72].

As a mechanism of UV-induced photooxidative damage, silymarin has shown the ability to modulate the activation of transcription factors NF-κB and AP-1 in HaCaT keratinocytes [73]. Further, silymarin treatment prevents UVB-induced immune suppression and oxidative stress in vivo: silymarin treatment of C3H/HeN mice inhibited UVB-induced suppression of contact hypersensitivity, inhibition of infiltrating leukocytes and myeloperoxidase activity [74].

10.5.3 Resveratrol

Resveratrol (*trans*-3,4′,5-trihydroxystilbene) is a polyphenolic phytoalexin found largely in the skin and seeds of grapes, nuts, fruits and red wine. Resveratrol is a potent antioxidant with antiinflammatory and antiproliferative properties [75]. Topical application of resveratrol to SKH-1 hairless mice prior to UVB irradiation results in significant inhibition of UVB-induced skin edema [76] and causes a significant decrease in UVB-mediated generation of hydrogen peroxide and infiltration of leukocytes. Pretreatment of NHEK with resveratrol inhibits UVB-mediated activation of the NF-κB pathway [77]. The protective effects of resveratrol against the damage induced by multiple UVB exposures occur via modulations in the cki-cyclin-cdk network and MAPK pathway [78].

10.5.4 Genistein

Soybeans are a rich source of the flavonoids isoflavones, the most potent being genistein and daidzein. Genistein (4′,5,7-trihydroxyisoflavone) is primarily present in soy, in ginkgo biloba extract, Greek oregano and Greek sage. Genistein has been shown to possess antioxidant and anticarcinogenic effects in skin [79]. Genistein significantly enhanced the activities

of antioxidant enzymes in the skin of SENCAR mice [80]. Treatment of the human keratinocyte cell line NCTC 2544 with genistein prevents UVA-induced enhancement of the DNA-binding activity of signal transducer and activator of transcription-1 (STAT-1) by acting as a tyrosine kinase inhibitor, thus limiting lipid peroxidation and increases in ROS formation [81]. Topical application of genistein to SENCAR mice before UVB irradiation reduces c-fos and c-jun expression in the skin in a dose-dependent manner [82]. Genistein is also able to significantly decrease the inflammatory edema reaction and suppress contact hypersensitivity induced by moderate doses of solar-simulated UV radiation [83]. Treatment of animal skin with genistein prior to UVB exposure resulted in significant inhibition of UVB-induced H_2O_2, malondialdehyde and 8-OH-dG production [84].

References

1. Sies H (1986) Biochemistry of oxidative stress. Angew Chem Int Ed Engl 25:1058–1071
2. Sies H (1993) Strategies of antioxidant defense. Eur J Biochem 215:213–219
3. Hollman PCH, Katan MB (1999) Dietary flavonoids: intake, health effects and bioavailability. Food Chem Toxicol 37:937–942
4. Krutmann J (2000) Ultraviolet A radiation-induced biological effects in human skin: relevance for photoaging and photodermatosis. J Dermatol Sci 23:S22–S26
5. Boelsma E, van de Vijver LP, Goldbohm RA, Klopping-Ketelaars IA, Hendriks HF, Roza L (2003) Human skin condition and its associations with nutrient concentrations in serum and diet. Am J Clin Nutr 77(2):348–355
6. Sies H, Stahl W (2004) Nutritional protection against skin damage from sunlight. Annu Rev Nutr 24:173–200
7. Sies H, Stahl W (2004) Carotenoids and UV protection. Photochem Photobiol Sci 3(8):749–752
8. Klotz LO, Holbrook NJ, Sies H (2001) UVA and singlet oxygen as inducers of cutaneous signaling events. Curr Probl Dermatol 29:95–113
9. Wenk J, Brenneisen P, Meewes C, Wlaschek M, Peters T, Blaudschun R, et al (2001) UV-induced oxidative stress and photoaging. Curr Probl Dermatol 29:83–94
10. Wlaschek M, Tantcheva-Poor I, Naderi L, Ma W, Schneider LA, Razi-Wolf Z, et al (2001) Solar UV

irradiation and dermal photoaging. J Photochem Photobiol B 63(1-3):41–51

11. Demmig-Adams B, Adams WW III (2002) Antioxidants in photosynthesis and human nutrition. Science 298(5601):2149–2153

12. Godar DE, Urbach F, Gasparro FP, van der Leun JC (2003) UV doses of young adults. Photochem Photobiol 77(4):453–457

13. Godar DE (2001) UV doses of American children and adolescents. Photochem Photobiol 74(6):787–793

14. Godar DE, Wengraitis SP, Shreffler J, Sliney DH (2001) UV doses of Americans. Photochem Photobiol 73(6):621–629

15. Black HS, Rhodes LE (2001) Systemic photoprotection: dietary intervention and therapy. In: Giacomoni PU (ed) Sun protection in man. Elsevier, Amsterdam, pp 573–591

16. Stahl W, van den Berg H, Arthur J, Bast A, Dainty J, Faulks RM, et al (2002) Bioavailability and metabolism. Mol Aspects Med 23(1-3):39–100

17. Traber MG, Sies H (1996) Vitamin E in humans: demand and delivery. Annu Rev Nutr 16:321–347

18. Yeum KJ, Russell RM (2002) Carotenoid bioavailability and bioconversion. Annu Rev Nutr 22:483–504

19. Omenn GS, Goodman GE, Thornquist MD, Balmes J, Cullen MR, Glass A, et al (1996) Risk factors for lung cancer and for intervention effects in CARET, the beta-carotene and retinol efficacy trial. J Natl Cancer Inst 88:1550–1559

20. Albanes D, Heinonen OP, Taylor PR, Virtamo J, Edwards BK, Rautalahti M, et al (1996) Alpha-tocopherol and beta-carotene supplements and lung cancer incidence in the alpha-tocopherol, beta-carotene cancer prevention study: effects of base-line characteristics and study compliance. J Natl Cancer Inst 88(21):1560–1570

21. Traber MG, Serbinova EA, Packer L (1999) Biological activities of tocotrienols and tocopherols. In: Packer L, Hiramatsu M, Yoshikawa T (eds) Antioxidant food supplements in human health. Academic Press, San Diego, pp 55–71

22. Brigelius-Flohe R, Traber MG (1999) Vitamin E: function and metabolism. FASEB J 13(10):1145–1155

23. Liebler DC (1993) The role of metabolism in the antioxidant function of vitamin E. Crit Rev Toxicol 23:147–169

24. Weber P, Bendich A, Schalch W (1996) Vitamin C and human health – a review of recent data relevant to human requirements. Int J Vitam Nutr Res 66:19–30

25. McCall MR, Frei B (1999) Can antioxidant vitamins materially reduce oxidative damage in humans? Free Radic Biol Med 26:1034–1053

26. Fuchs J, Kern H (1998) Modulation of UV-light-induced skin inflammation by D-alpha-tocopherol and L-ascorbic acid: a clinical study using solar simulated radiation. Free Radic Biol Med 25:1006–1012

27. Eberlein-König B, Placzek M, Przybilla B (1998) Protective effect against sunburn of combined systemic ascorbic acid (vitamin C) and D-alpha-tocopherol (vitamin E). J Am Acad Dermatol 38:45–48

28. Stahl W, Heinrich U, Jungmann H, Sies H, Tronnier H (2000) Carotenoids and carotenoids plus vitamin E protect against ultraviolet light-induced erythema in humans. Am J Clin Nutr 71:795–798

29. Werninghaus K, Meydani M, Bhawan J, Margolis R, Blumberg JB, Gilchrest BA (1994) Evaluation of the photoprotective effect of oral vitamin E supplementation. Arch Dermatol 130:1257–1261

30. Fuchs J, Weber S, Podda M, Groth N, Herrling T, Packer L, et al (2003) HPLC analysis of vitamin E isoforms in human epidermis: correlation with minimal erythema dose and free radical scavenging activity. Free Radic Biol Med 34(3):330–336

31. Thiele JJ, Weber SU, Packer L (1999) Sebaceous gland secretion is a major physiologic route of vitamin E delivery to skin. J Invest Dermatol 113(6):1006–1010

32. Thiele JJ, Schroeter C, Hsieh SN, Podda M, Packer L (2001) The antioxidant network of the stratum corneum. Curr Probl Dermatol 29:26–42

33. Thiele JJ (2001) Oxidative targets in the stratum corneum. A new basis for antioxidative strategies. Skin Pharmacol Appl Skin Physiol 14:87–91

34. McArdle F, Rhodes LE, Parslew R, Jack CI, Friedmann PS, Jackson MJ (2002) UVR-induced oxidative stress in human skin in vivo: effects of oral vitamin C supplementation. Free Radic Biol Med 33(10):1355–1362

35. Lin JY, Selim MA, Shea CR, Grichnik JM, Omar MM, Monteiro-Riviere NA, et al (2003) UV photoprotection by combination topical antioxidants vitamin C and vitamin E. J Am Acad Dermatol 48(6):866–874

36. Dreher F, Gabard B, Schwindt DA, Maibach HI (1998) Topical melatonin in combination with vitamins E and C protects skin from ultraviolet-induced erythema: a human study in vivo. Br J Dermatol 139(2):332–339

37. Fuchs J (1998) Potentials and limitations of the natural antioxidants RRR-alpha-tocopherol, L-ascorbic acid and beta-carotene in cutaneous photoprotection. Free Radic Biol Med 25:848–873

38. McVean M, Liebler DC (1999) Prevention of DNA photodamage by vitamin E compounds and sunscreens: roles of ultraviolet absorbance and cellular uptake. Mol Carcinog 24(3):169–176

39. Thiele J, Dreher F, Packer L (2000) Antioxidant defense systems in skin. In: Elsner P, Maibach H (eds)

Cosmeceuticals. Marcel Dekker, New York, pp 145–187

40. F'guyer S, Afaq F, Mukhtar H (2003) Photochemoprevention of skin cancer by botanical agents. Photodermatol Photoimmunol Photomed 19(2):56–72

41. Packer L, Obermüller-Jevic U, Kraemer K, Sies H (eds) (2005) Carotenoids and retinoids – molecular aspects and health issues. AOCS Press, Champaign

42. Krinsky NI, Mayne ST, Sies H (eds) (2005) Carotenoids in health and disease. Marcel Dekker, New York

43. Peng Y-M, Peng Y-S, Lin Y (1993) A nonsaponification method for the determination of carotenoids, retinoids, and tocopherols in solid human tissues. Cancer Epidemiol Biomark Prev 2:139–144

44. Stahl W, Heinrich U, Jungmann H, von Laar J, Schietzel M, Sies H, et al (1998) Increased dermal carotenoid levels assessed by noninvasive reflection spectrophotometry correlate with serum levels in women ingesting Betatene. J Nutr 128:903–907

45. Gollnick HPM, Hopfenmüller W, Hemmes C, Chun SC, Schmid C, Sundermeier K, et al (1996) Systemic beta carotene plus topical UV-sunscreen are an optimal protection against harmful effects of natural UV-sunlight: results of the Berlin-Eilath study. Eur J Dermatol 6:200–205

46. Heinrich U, Gärtner C, Wiebusch M, Eichler O, Sies H, Tronnier H, et al (2003) Supplementation with beta-carotene or a similar amount of mixed carotenoids protects humans from UV-induced erythema. J Nutr 133(1):98–101

47. Lee J, Jiang S, Levine N, Watson RR (2000) Carotenoid supplementation reduces erythema in human skin after simulated solar radiation exposure. Proc Soc Exp Biol Med 223:170–174

48. Mathews-Roth MM, Pathak MA, Parrish JA, Fitzpatrick TB, Kass EH, Toda K, et al (1972) A clinical trial of the effects of oral beta-carotene on the responses of human skin to solar radiation. J Invest Dermatol 59:349–353

49. Garmyn M, Ribaya-Mercado JD, Russell RM, Bhawan J, Gilchrest BA (1995) Effect of beta-carotene supplementation on the human sunburn reaction. Exp Dermatol 4:104–111

50. Wolf C, Steiner A, Hönigsmann H (1988) Do oral carotenoids protect human skin against ultraviolet erythema, psoralen phototoxicity, and ultraviolet-induced DNA damage? J Invest Dermatol 90:55–57

51. McArdle F, Rhodes LE, Parslew RA, Close GL, Jack CI, Friedmann PS, et al (2004) Effects of oral vitamin E and beta-carotene supplementation on ultraviolet radiation-induced oxidative stress in human skin. Am J Clin Nutr 80(5):1270–1275

52. Biesalski HK, Obermüller-Jevic UC (2001) UV light, beta-carotene and human skin – beneficial and potentially harmful effects. Arch Biochem Biophys 389(1):1–6

53. The Alpha-Tocopherol, Beta Carotene Cancer Prevention Study Group (1994) The effect of vitamin E and beta carotene on the incidence of lung cancer and other cancers in male smokers. N Engl J Med 330:1029–1035

54. Pelz R, Schmidt-Faber B, Heseker H (1998) Carotenoid intake in the German National Food Consumption Survey (in German). Z Ernährungswiss 37(4):319–327

55. Gärtner C, Stahl W, Sies H (1997) Lycopene is more bioavailable from tomato paste than from fresh tomatoes. Am J Clin Nutr 66:116–122

56. Stahl W, Heinrich U, Wiseman S, Eichler O, Sies H, Tronnier H (2001) Dietary tomato paste protects against ultraviolet light-induced erythema in humans. J Nutr 131(5):1449–1451

57. Aust O, Stahl W, Sies H, Tronnier H, Heinrich U (2005) Supplementation with tomato-based products increases lycopene, phytofluene, and phytoene levels in human serum and protects against UV-light-induced erythema. Int J Vitam Nutr Res 75(1):54–60

58. Dell'Aica I, Dona M, Sartor L, Pezzato E, Garbisa S (2002) (−)-Epigallocatechin-3-gallate directly inhibits MT1-MMP activity, leading to accumulation of nonactivated MMP-2 at the cell surface. Lab Invest 82(12):1685–1693

59. Vayalil PK, Mittal A, Hara Y, Elmets CA, Katiyar SK (2004) Green tea polyphenols prevent ultraviolet light-induced oxidative damage and matrix metalloproteinase expression in mouse skin. J Invest Dermatol 122(6):1480–1487

60. Rittie L, Fisher GJ (2002) UV-light-induced signal cascades and skin aging. Ageing Res Rev 1(4):705–720

61. Kim J, Hwang JS, Cho YK, Han Y, Jeon YJ, Yang KH (2001) Protective effects of (−)-epigallocatechin-3-gallate on UVA- and UVB-induced skin damage. Skin Pharmacol Appl Skin Physiol 14(1):11–19

62. Song DU, Jung YD, Chay KO, Chung MA, Lee KH, Yang SY, et al (2002) Effect of drinking green tea on age-associated accumulation of Maillard-type fluorescence and carbonyl groups in rat aortic and skin collagen. Arch Biochem Biophys 397(2):424–429

63. Katiyar SK, Mukhtar H (2001) Green tea polyphenol (−)-epigallocatechin-3-gallate treatment to mouse skin prevents UVB-induced infiltration of leukocytes, depletion of antigen-presenting cells, and oxidative stress. J Leukoc Biol 69(5):719–726

64. Katiyar SK, Challa A, McCormick TS, Cooper KD, Mukhtar H (1999) Prevention of UVB-induced immunosuppression in mice by the green tea polyphenol (−)-epigallocatechin-3-gallate may be associated with alterations in IL-10 and IL-12 production. Carcinogenesis 20(11):2117–2124

65. Katiyar SK, Afaq F, Azizuddin K, Mukhtar H (2001) Inhibition of UVB-induced oxidative stress-medi-

ated phosphorylation of mitogen-activated protein kinase signaling pathways in cultured human epidermal keratinocytes by green tea polyphenol (–)-epigallocatechin-3-gallate. Toxicol Appl Pharmacol 176(2):110–117

66. Afaq F, Adhami VM, Ahmad N, Mukhtar H (2003) Inhibition of ultraviolet B-mediated activation of nuclear factor kappaB in normal human epidermal keratinocytes by green tea constituent (–)-epigallocatechin-3-gallate. Oncogene 22(7):1035–1044

67. Barthelman M, Bair WB III, Stickland KK, Chen W, Timmermann BN, Valcic S, et al (1998) (–)-Epigallocatechin-3-gallate inhibition of ultraviolet B-induced AP-1 activity. Carcinogenesis 19(12):2201–2204

68. Chen W, Dong Z, Valcic S, Timmermann BN, Bowden GT (1999) Inhibition of ultraviolet B-induced c-fos gene expression and p38 mitogen-activated protein kinase activation by (–)-epigallocatechin gallate in a human keratinocyte cell line. Mol Carcinog 24(2):79–84

69. Afaq F, Ahmad N, Mukhtar H (2003) Suppression of UVB-induced phosphorylation of mitogen-activated protein kinases and nuclear factor kappa B by green tea polyphenol in SKH-1 hairless mice. Oncogene 22(58):9254–9264

70. Uehara M, Sugiura H, Sakurai K (2001) A trial of oolong tea in the management of recalcitrant atopic dermatitis. Arch Dermatol 137(1):42–43

71. Wagner H, Diesel P, Seitz M (1974) The chemistry and analysis of silymarin from Silybum marianum Gaertn (in German). Arzneimittelforschung 24(4):466–471

72. Katiyar SK, Korman NJ, Mukhtar H, Agarwal R (1997) Protective effects of silymarin against photocarcinogenesis in a mouse skin model. J Natl Cancer Inst 89(8):556–566

73. Dhanalakshmi S, Mallikarjuna GU, Singh RP, Agarwal R (2004) Dual efficacy of silibinin in protecting or enhancing ultraviolet B radiation-caused apoptosis in HaCaT human immortalized keratinocytes. Carcinogenesis 25(1):99–106

74. Katiyar SK (2002) Treatment of silymarin, a plant flavonoid, prevents ultraviolet light-induced immune suppression and oxidative stress in mouse skin. Int J Oncol 21(6):1213–1222

75. Tsai SH, Lin-Shiau SY, Lin JK (1999) Suppression of nitric oxide synthase and the down-regulation of the activation of NFkappaB in macrophages by resveratrol. Br J Pharmacol 126(3):673–680

76. Afaq F, Adhami VM, Ahmad N (2003) Prevention of short-term ultraviolet B radiation-mediated damages by resveratrol in SKH-1 hairless mice. Toxicol Appl Pharmacol 186(1):28–37

77. Adhami VM, Afaq F, Ahmad N (2003) Suppression of ultraviolet B exposure-mediated activation of NF-kappaB in normal human keratinocytes by resveratrol. Neoplasia 5(1):74–82

78. Reagan-Shaw S, Afaq F, Aziz MH, Ahmad N (2004) Modulations of critical cell cycle regulatory events during chemoprevention of ultraviolet B-mediated responses by resveratrol in SKH-1 hairless mouse skin. Oncogene 23(30):5151–5160

79. Wei H, Bowen R, Cai Q, Barnes S, Wang Y (1995) Antioxidant and antipromotional effects of the soybean isoflavone genistein. Proc Soc Exp Biol Med 208(1):124–130

80. Cai Q, Wei H (1996) Effect of dietary genistein on antioxidant enzyme activities in SENCAR mice. Nutr Cancer 25(1):1–7

81. Maziere C, Dantin F, Dubois F, Santus R, Maziere J (2000) Biphasic effect of UVA radiation on STAT1 activity and tyrosine phosphorylation in cultured human keratinocytes. Free Radic Biol Med 28(9):1430–1437

82. Wang Y, Zhang X, Lebwohl M, DeLeo V, Wei H (1998) Inhibition of ultraviolet B (UVB)-induced c-fos and c-jun expression in vivo by a tyrosine kinase inhibitor genistein. Carcinogenesis 19(4):649–654

83. Widyarini S, Spinks N, Husband AJ, Reeve VE (2001) Isoflavonoid compounds from red clover (Trifolium pratense) protect from inflammation and immune suppression induced by UV radiation. Photochem Photobiol 74(3):465–470

84. Wei H, Cai Q, Tian L, Lebwohl M (1998) Tamoxifen reduces endogenous and UV light-induced oxidative damage to DNA, lipid and protein in vitro and in vivo. Carcinogenesis 19(6):1013–1018

Chapter 11

Hormone Replacement Therapy and Skin Aging

11

Glenda K. Hall, Tania J. Phillips

Contents

11.1 Introduction

Aging is a process associated with a decline in the function of all organ systems. The skin is the largest organ of the human body and is notably affected by aging. Cutaneous aging is characterized by atrophy, wrinkling, dryness, increased laxity, and poor wound healing (Table 11.1). It is influenced by several factors including genetics, cumulative sun exposure, and hormonal status [1]. Declining levels of hormones, especially estrogen, are associated with the aging process. While the effects of estrogen on skin function are not fully understood, effects of estrogen supplementation upon some of these age-related skin conditions in the elderly, specifically postmenopausal women, have been reported. Postmenopausal women who utilize estrogen supplementation tend to have lower wrinkle scores, less xerosis, and relatively thicker skin than women not receiving hormone replacement therapy (HRT). Estrogen supplementation appears to improve wound healing in elderly women.

Table 11.1. Skin functions known to decline with age

Epidermal turnover
Wound healing
Barrier function
Immune function
Sweat production
Sebum production
Vitamin D production

11.2 Estrogens and the Menopause

The estrogens, estradiol, estrone and estriol, are C-18 steroids differentiated by the C-17 side chain [2]. Their individual potencies as estrogens vary, with estradiol being the most potent and estriol being the least potent [2, 3]. Cholesterol is the parent steroid from which all gonadal steroids are derived [4]. Estradiol is predominantly synthesized in the granulosa cells of the ovary where it is converted from androstenedione produced in the theca cells. Estrone is the product of peripheral aromatization and the metabolism of estradiol and estrone results in the formation of estriol [2, 3, 5].

The skin is the largest non-reproductive organ targeted by estrogen. Estrogen interacts with the skin and other tissues through receptors. Two estrogen receptors (ER) have been identified: ERα and ERβ [6, 7]. Both ERs have 60% homology and nearly equal binding affinity for a large number of ligands [8–11]. However, each ER has variable expression among different tissue types as well as within the skin. The vaginal epithelium has the largest concentration of ER within the genital tract [12]. Skin on the face expresses larger concentrations of ER than breast or thigh skin [13]. Despite having structural and functional similarities, the variable concentration of these receptors within the skin suggests that each has a different, cell-specific role [14]. With the menopause, the expression of ER declines [12, 15].

Estrogen production varies with age and is regulated by the hypothalamic-pituitary axis (HPA) (Fig. 11.1). The pituitary gland, under the pulsatile stimulus of hypothalamic gonadotropin-releasing hormone, secretes luteinizing hormone (LH) and follicle-stimulating hormone (FSH). LH stimulates the theca cells to produce androstenedione while FSH stimulates the granulosa cells to convert androstenedione to estradiol. The resulting increase in serum estradiol exerts negative feedback upon the pituitary secretions thereby maintaining serum estradiol levels in the 10–20 mIU/ml range throughout adulthood [16]. At birth, serum estradiol levels drop precipitately and remain low until the onset of puberty at which time a steady rise occurs in women until maturity is

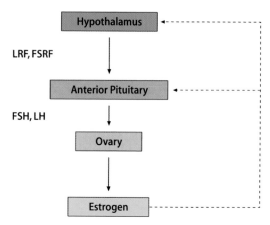

Fig. 11.1. Feedback control of the anterior pituitary and ovary. *LFR* luteotropin, *FSRF* follicle-stimulating releasing factor, *LH* lutenizing hormone, *FSH* follicle-stimulating hormone (reprinted from Phillips TJ, Demircay Z, Sahu M. Hormonal effects on skin aging. Clin Geriatr Med 2001;17:662; with permission from Elsevier)

reached. Throughout female adulthood, serum estradiol exhibits a diurnal rhythm correlated to LH stimulation [17]. At a genetically predetermined time in a woman's life, menopause occurs and menstruation ends [18]. During the premenopausal years, FSH and LH gradually rise due to a decline in estradiol production resulting in a loss of negative feedback to the pituitary gland [16]. With minimal estradiol production, estrone becomes the predominant circulating estrogen due to continued peripheral aromatization [19, 20].

With the menopause, the decrease in circulating estradiol, along with decreases in the expression of ER, may lead to the various physiological changes observed during this time period.

11.2.1 Synopsis

Estradiol, estrone, and estriol (listed in order of decreasing potency) interact with the skin through α and β ER whose concentration varies within the skin and decreases during the menopause.

The hypothalamic-pituitary axis regulates the production of estrogen which varies with

Table 11.2. Summary of the HERS and WHI trials (*CCE* continuous conjugated estrogens, *MPA* medroxyprogesterone acetate, *WHI* Womens Health Initiative, *HERS* Heart and Estrogen/Progestin Replacement Study)

HERS trial [26]	2,763 postmenopausal women less than 80 years of age with known coronary disease were randomized to either estrogen plus progestin CEE 0.625 mg/day plus MPA 2.5 mg/day (*n*=1,380) or placebo (*n*=1,383)	After 4.1 years of follow-up, no overall cardiovascular benefit and an early increase in coronary events was found with the use of estrogen and progestin
WHI trial [25]	16,608 healthy postmenopausal women aged 50–79 years were randomized to either estrogen plus progestin 0.625 mg CEE plus 2.5 mg MPA (*n*=8,506) or placebo (*n*=8,102)	Stopped early after 5.2 years of follow-up (expected follow-up 8.5 years) when the global index indicated that risks of HRT exceed benefits. Women treated with CEE/MPA had 1.26-fold greater risk of breast cancer than placebo group

age and declines during the perimenopausal years.

11.3 Hormone Replacement Therapy and Aging

HRT was first introduced 70 years ago and, in 2000, was the second most commonly prescribed drug [21]. Relief of menopausal symptoms and either treatment or prevention of osteoporosis are the main reasons documented for prescribing HRT [22]. Previously estrogen was also considered beneficial in the prevention of cardiovascular disease [23]. Recently, however, women have been forced to reconsider the benefits of HRT. While the risks seem low [24], findings from the Women's Health Initiative Trial (WHI) and the HERS (Heart and Estrogen/Progestin Replacement Study) trial have demonstrated a small but apparent increased risk of breast cancer and coronary artery disease in those women utilizing HRT [25, 26] (Table 11.2). However, these studies were performed in older, predominantly obese women, many of whom had preexisting cardiovascular risk factors, such as hypertension and dyslipidemias. In addition, these studies used specific HRT regimens, and the results cannot necessarily be extrapolated to general HRT use [27].

The decline in skin appearance from the perimenopausal years onward suggests that the decrease in sex steroids may play a role in aging. The most common changes include skin atrophy, dryness, laxity, wrinkling, and poor wound healing. While several studies have suggested that many of these signs can be improved or reversed with the use of estrogen, the evidence remains controversial. The decision to use supplemental estrogen for the treatment of cutaneous aging is further clouded by the recent negative associations linked with estrogen use. Thus, physicians should allow women to make informed decisions regarding the safety and efficacy of supplemental estrogen.

The issues surrounding HRT are complicated. Women should carefully consider the risks and benefits of HRT in consultation with their physician (Table 11.3). In general, short-term HRT is indicated for the relief of postmenopausal symptoms, but it would seem prudent to avoid long-term HRT for the prevention of disease [24, 27].

11.3.1 Dermal Collagen Content

Thinning of the skin, clinically appreciated by easy tearing and bruising, is a common sign of aging. Several metabolic activities decrease with aging and thinning of the skin is believed to result from a decrease in dermal collagen synthesis [28, 29]. An average decline of 1–2% per year in dermal collagen content following the menopause has been reported [30]. The association between collagen loss and postmenopausal age, rather than chronological age, may reflect a hormonal etiology [28, 29]. Thus, several stud-

Table 11.3. Risks and benefits of postmenopausal HRT

		Reference
Benefits	Relief of menopausal symptoms	67
	Relief of urogenital atrophy and its symptoms (topical estrogen)	68
	Prevention of fractures in women with osteoporosis	69
Risks	Increased risk of breast cancer in women who used HRT for over 5 years	69
	No role in secondary cardiovascular disease (CVD) prevention	27
	Controversial role in primary CVD prevention	27
	Stroke in older women (17β-estradiol)	70
	Venous thromboembolism	25–27

ies have evaluated the effectiveness of estrogen supplementation on increasing skin thickness and dermal collagen content. The results are quite varied, and comparison among the studies is difficult due to variations in the study designs including the route of estrogen administration, the use of combined progestins, and the length of estrogen exposure. Nevertheless, the majority of evidence demonstrates an increase in skin thickness and/or dermal collagen content with postmenopausal estrogen supplementation [31–37]. Interestingly, the effect of estrogen appears to be dependent on baseline collagen content at the onset of treatment. Estrogen has a preventative role in collagen loss in those women with higher initial collagen levels, and a therapeutic role for collagen synthesis in those women with lower initial collagen levels [38, 39].

Initial studies performed throughout the 1980s by Brincat and various collaborators demonstrated increases in skin thickness and/or collagen content with several different hormone replacement regimens, including varying doses of testosterone [29, 30, 38–40]. An ideal estrogen dose was postulated, based on suboptimal increases, or even decreases, observed in dermal collagen content with varying estrogen doses [38]. More recently, randomized placebo-controlled studies have assessed the effects of estrogen on skin thickness and dermal collagen content. In 1994, a randomized placebo-controlled study of 60 postmenopausal nuns reported increased skin thickness at the level of the greater trochanter after 12 months of therapy with 0.625 mg conjugated estrogens [31].

Skin thickness measurements were performed with ultrasound. Skin biopsies obtained from the same location revealed a 33% increase in dermal thickness. The patient population was selected to decrease confounding variables of aging such as smoking and extrinsic photoaging due to extensive sun exposure. Another randomized, placebo-controlled study evaluated 118 postmenopausal women on one of four possible treatment regimens over a 12-month period [32]. Women were randomized to 0.625 mg conjugated estrogens in a 25-day cycle, 0.625 mg conjugated estrogens every day, 50 μg transdermal estradiol in a 24-day cycle, or no treatment. Each active treatment group received medroxyprogesterone for the last 12 days of each cycle. All active treatment groups had increases in the dermal collagen level while the placebo group experienced a 3.2% decline in dermal collagen content. Transdermal estradiol in combination with a progestin demonstrated the greatest efficacy, with a 5.1% increase in collagen, when compared to oral conjugated estrogens [32]. More recently, a randomized double-blind placebo-controlled study of estrogen in 41 postmenopausal women revealed a 6.5% increase in dermal collagen content for the active treatment group and no change in dermal collagen content for the placebo group [33]. Skin biopsies were obtained from the medial upper arm and were analyzed using computerized image analysis. The active treatment group received 2 mg estradiol in a 21-day cycle with 1 mg cyproterone for the last 12 days of each cycle. Treatment was continued for 6 months.

11

Several open-label studies have supported the findings from the randomized, controlled trials. A study of 100 postmenopausal women treated with either combination HRT (estradiol and cyproterone) or calcium carbonate for 6 months demonstrated increased skin thickness on ultrasound in the active treatment group and a significant decrease in the calcium treated group [34]. A study of topical estradiol gel confirmed the localized effect of topical estrogen on the skin [35]. The gel was applied to one-half of the lower abdomen for 3 months in 12 postmenopausal women and a vehicle-only gel was applied to the contralateral half of the lower abdomen to serve as a control site. A 38% increase from baseline in skin hydroxyproline content on the estrogen-treated side was demonstrated. In addition, blister fluid analysis revealed increased propeptides of collagen on the estrogen-treated side indicating a stimulatory effect on collagen synthesis. Both local and systemic effects of topical estrogen supplementation were demonstrated in a study of 98 postmenopausal women who were divided into two equally-numbered groups based on hormone replacement utilization [36]. The active treatment group received either estradiol gel ($n=36$) or transdermal estradiol ($n=13$). Skin thickness measurements at five locations (inner and outer forearm, forehead, breast, and estrogen application site) were obtained utilizing B-mode ultrasound high-resolution echography. Statistically significant increases in skin thickness as compared to control were found at the inner and outer forearms, and highly significant increases were demonstrated at the breast and the site of estrogen application.

Some studies have disputed the effects of estrogen supplementation on skin thickness and collagen content. Bolognia et al. performed a double-blind placebo-controlled study of 46 postmenopausal women in whom cutaneous symptoms, including the complaint of thinning skin, and cutaneous signs, including bruising, were assessed. No statistically significant difference in the complaints of thinning skin or clinical sign of bruising was found in the group treated with 6 months of 17β-estradiol compared to the placebo group [37]. An open non-randomized trial of 43 postmenopausal women showed no effects of treatment with either 2 mg 17β-estradiol or 2 mg estradiol valerate after 12 months [41, 42]. Utilizing four independent methods to detect changes in the connective tissue (ultrasonographic measurement of skin thickness, assessment of total collagen with a colorimetric method, determination of de novo synthesis of collagen by measuring procollagen propeptides in blister fluid, and immunohistochemistry), the authors concluded that 1 year of systemic estrogen did not affect skin thickness, the amount of collagen, or the rate of collagen synthesis. Interestingly, both of these studies enrolled women with a relatively young postmenopausal age. Bolognia et al. required amenorrhea for 4 months while the latter study required amenorrhea for at least 6 months but less than 2 years. As discussed above, a young postmenopausal age may be associated with a higher baseline collagen content, thus introducing selection bias into the data.

In summary, HRT has been shown to increase skin thickness and dermal collagen content in several studies. However, the extent of the increase varies depending on the dose, route of administration, and duration of treatment. In addition, some studies have not shown positive effects of HRT.

11.3.1.1 Synopsis

Age-related atrophy of the skin has been correlated with decreases in dermal collagen which occurs at a rate of 1–2% per year after the menopause.

The efficacy of supplemental estrogen in treating skin atrophy may be related to the baseline collagen content at the onset of treatment. Thus, it is prophylactic in early menopause and therapeutic later on.

Not all studies have demonstrated beneficial effects of estrogen supplementation on skin atrophy.

11.3.2 Dryness: Water-Holding Capacity and Epidermal Lipid Layer

In a clinical examination of 3875 postmenopausal women, the utilization of HRT decreased the likelihood of dry skin when compared to those who were not receiving HRT [43]. In another study, the use of 0.01% estradiol or 0.3% estriol for 6 months resulted in increased skin moisture via corneometry in 59 postmenopausal women. However, these increases did not reach statistical significance [44]. Skin hydration is highly influenced by the skin's ability to retain water. The status of the stratum corneum and the dermal content of glycosaminoglycans affects the skin's water-holding capacity. A study of 30 postmenopausal women receiving transdermal estrogen supplementation found increases in the water-holding capacity of the stratum corneum through measurements of transepidermal water loss [45]. Glycosaminoglycans are hydrophilic molecules that draw moisture into the dermis. Decreased dermal glycosaminoglycans are associated with aging [46] and are thus felt to contribute to dry skin, wrinkling, and atrophy [47]. Estrogen supplementation in animals has been shown to result in marked increases of glycosaminoglycans within 2 weeks of therapy [48], and one human study has demonstrated increased dermal hydroscopic qualities during states of elevated endogenous estrogen (pregnancy), suggesting a similar mechanism [49].

Changes in the lipid layer of the epidermis are frequently associated with aging and may affect the water-holding capacity of the skin. Significant variation in stratum corneum sphingolipids has been noted among women of varying ages thus suggesting a possible hormonal influence [50]. Increased skin surface lipids have been measured, utilizing a sebumeter, in postmenopausal women receiving transdermal estradiol when compared to non-treated controls (total $n=98$) [36]. Another study also found increased skin surface lipids and skin moisture in 15 women treated with 6 months of combination HRT (estradiol plus progestin) [51]. These findings suggests that postmenopausal supplementation of estrogen may enhance the skin barrier function, thus preventing dryness.

11.3.2.1 Synopsis

The use of HRT is correlated with a decreased likelihood of dry skin. This finding may be related to increases in hydrophilic dermal glycosaminoglycans.

The composition of stratum corneum sphingolipids is hormonally influenced. Increases in these skin surface lipids, as seen with HRT, may result in improved skin hydration and barrier function.

11.3.3 Wrinkling and Laxity: Elastic Fiber Content

Laxity and wrinkling are cutaneous signs of aging related to the loss of skin elasticity. Estrogen supplementation has been demonstrated to improve non-invasive measurements of laxity and wrinkling. However, studies evaluating the effects of estrogens on the elastic components of dermal tissue have been controversial. In a study of 180 women aged 18–67 years, a progressive increase in skin extensibility and a decrease in skin elasticity was observed with aging [52]. In a sub-cohort of 30 postmenopausal women, the use of HRT (either 0.625 mg/day conjugated estrogens or 2 mg/day estradiol, each with an associated progestin for the last 12 days of each cycle) slowed the progression of these skin changes when compared to age-matched controls not utilizing HRT. Supporting evidence demonstrated steep increases in skin extensibility, measured with computerized suction devices, in menopausal women not utilizing HRT when compared to those utilizing HRT [53].

Studies assessing wrinkling have also been promising. Decreased wrinkle measurements via profilometry were reported in 59 preclimacteric women who applied either estradiol or estriol to the face for 6 months [44]. A randomized, double-blind, placebo controlled study of 54 postmenopausal women reported an improvement in fine wrinkling after applying Premarin cream to the face at bed-time for 24 weeks [54]. However, subject self-assessments did not reveal any changes in perception of facial appearance including fine wrinkling. Data from the largest population-based study to date, the

11

First National Health and Nutrition Examination which included 28,000 civilians and 3,875 postmenopausal women, indicated that the use of estrogen is associated with less wrinkling in postmenopausal women. Reliability of estrogen use among the 3,875 postmenopausal women, however, has been questioned given that the history of estrogen use was obtained in follow-up surveys conducted 9–15 years after the initial assessments [43].

Invasive studies evaluating the effects of estrogens on elastic fibers have not been uniformly conclusive. Bolognia et al. found degenerative changes in the dermal elastic fibers of women with premature menopause and suggested a possible relationship between estrogen deprivation and changes in dermal elastic fibers [37]. Two small studies have demonstrated increased elastic fibers after therapy with estrogen supplementation, although larger studies have not reached the same conclusion. Localized increases in the concentration and size of elastic fibers were found in 7 of 14 postmenopausal women applying 2 mg of estriol to the abdomen daily for 3 weeks [55]. Morphological improvements in elastic fibers at the site of estradiol application after 3 months of therapy were found in a study of 12 postmenopausal women [35]. On the contrary, an open study of 43 postmenopausal women did not find any change in the proportional area of elastic fibers assessed by computerized image analysis of light microscopy images after 12 months therapy with estradiol [41, 42]. Finally, a recent randomized double-blind placebo-controlled trial of 41 postmenopausal women, despite showing improvements in collagen content, did not show any changes in elastic fiber content after 6 months estradiol therapy [33].

While dermatological examinations have shown clinically significant improvements in laxity and wrinkling after the use of supplemental estrogen, patient self-assessments have not always corroborated these findings.

11.3.3.1 Synopsis

Extensibility of the skin progresses with age. The use of HRT has been shown to slow this progression when compared to cohorts not utilizing HRT.

Fine wrinkling may also be decreased with the use of HRT, especially when applied to the face. However, the improvement does not necessarily correlate with improved patient satisfaction.

11.3.4 Wound Healing

Poor wound healing accompanies aging and is linked to an exuberant inflammatory response resulting in excessive proteolysis of structural elements important in keratinocyte migration [56, 57]. High neutrophil counts in chronic wounds result in elevated levels of proteolytic enzymes such as elastase and other matrix matalloproteinases (MMPs). Studies have demonstrated elevated levels of MMPs in the chronic wounds of elderly patients with reduced levels of fibronectin [58, 59]. Studies have also shown that estrogen plays a crucial role in wound healing [60, 61]. Improved wound healing of acute wounds has been demonstrated in elderly men and women treated with transdermal estradiol in comparison to placebo patch [62]. Other studies have demonstrated a decreased incidence of chronic wounds, such as venous ulcers and pressure ulcers, in women utilizing HRT [63, 64].

Several mechanisms to explain the effects of estrogen on wound healing have been proposed. An antiinflammatory mechanism is supported by a study that demonstrated decreased neutrophil chemotaxis, thereby decreased wound levels of elastase, with the use of estrogen. The decreased levels of elastase allow increased levels of fibronectin and thus improve cellular matrix formation and wound healing [62]. Transforming growth factor-β1 (TGF-β1), a cytokine involved in cell proliferation, differentiation, and matrix production, is also affected by estrogen status. TGF-β1 is decreased in the wounds of elderly females. Estrogen supplementation reverses that decrease and is associated with improved rates of wound healing [57]. Animal models have also demonstrated increased rates of wound repair with the administration of TGF-β1 [59].

Despite the beneficial effects of TGF-β1 on wound healing, this cytokine also adversely affects scarring profile (see below). Finally, studies evaluating macrophage migration inhibitory factor (MIF) have found it to be markedly elevated in the wounds of estrogen-deficient mice. In addition, mice devoid of the MIF gene do not demonstrate exuberant inflammation or delayed wound healing during states of estrogen deficiency [65]. This area of study is still under investigation.

Scarring, the end result of wound healing, is affected by hormonal status. Scars in older subjects are reported to have superior macroscopic (color, texture, and contour) and microscopic appearances when compared to the scars of younger individuals [59]. Favorable appearances include pale, flat scars with regenerated rete ridges, large papillary blood vessels, and a normal basket-weave organization of the collagen bundles. Changes in the quality of scarring may be related to TGF-β1. Neutralization of TGF-β1 in rodent cutaneous wounds led to anti-scarring effects [66]. Utilization of HRT in the elderly population, which is associated with increased levels of TGF-β1, has resulted in less favorable scarring profiles similar to younger subjects. This relationship suggests a hormonal and cytokine interaction on scar formation.

11.3.4.1 Synopsis

Chronic wound healing is characterized by an exuberant inflammatory response with high neutrophil counts and subsequent increases in proteolytic enzymes such as MMPs.

HRT and estrogen supplementation alone have been associated with improved wound healing in animal and human experimental models via an antiinflammatory mechanism and interaction with TGF-β1 and macrophage MIF.

11.4 Summary

Aging is an inevitable process that affects all organ systems including the skin. A decrease in serum estrogen accompanies aging and may contribute to age-related skin changes such as atrophy, wrinkling, dryness, and poor wound healing. Studies investigating the supplemental use of estrogen in the later years have demonstrated some beneficial effects on skin aging parameters. However, the use of HRT is controversial. Thus, the decision to use HRT should be made only after weighing the risks and benefits for each individual.

References

1. Yaar M, Gilchrest B (2003) Aging of skin. In: Freedberg IM, Eisen AZ, Wolff K, Austen KF, Goldsmith LA, Katz SI, Fitzpatrick TB (eds) Fitzpatrick's dermatology in general medicine, 5th edn, vol. II. McGraw-Hill, New York, pp 1386–1398
2. Carr BR (1998) Disorders of the ovaries and female reproductive tract. In: Wilson JD, Foster DW, Kronenberg HM, Larsen PR (eds) Williams textbook of endocrinology. WB Saunders, Philadelphia, pp 751–817
3. Williams CL, Stancel GM (1996) Estrogens and progestins. In: Hardman JG, Limbird LE, Molinoff PB, Rudden RW (eds) Goodman and Gilman's the pharmacological basis of therapeutics. McGraw-Hill, New York, pp 1411–1440
4. Barbieri R, Ryan K (1999) The menstrual cycle. In: Ryan K, Berkowitz R, Barbieri R, Dunaif A (eds) Kistner's gynecology and women's health, 7th edn. Mosby, St Louis, pp 32–34
5. Siiteri PK, MacDonald PC (1973) Role of extraglandular estrogen in human endocrinology. In: Greep RO, Astwood EB (eds) Handbook of physiology. Sect 7: Endocrinology, vol. II. American Physiological Society, Washington DC, pp 615–630
6. Greene GL, Gilna P, Waterfield M, et al (1986) Sequence and expression of human estrogen receptor complementary DNA Sci 231:1150–1154
7. Mosselman S, Polman J, Dijkema R (1996) ERβ: identification and characterization of a novel human estrogen receptor. FEBS Lett 392:49–53
8. Kuiper GG, Carlsson B, Grandien K, et al (1997) Comparison of the ligand binding specificity and transcript tissue distribution of estrogen receptor α and β. Endocrinology 138:863–870
9. Taylor AH, Al-Azzawi F (2000) Immunolocalization of oestrogen receptor beta in human tissues. J Mol Endocrinol 24:145–155
10. Saunders PTK, Maguire SM, Gaughan J, et al (1998) Expression of oestrogen receptor beta (Erbeta) in multiple rat tissues visualized by immunohistochemistry. J Endocrinol 154:R13–R16

11. MacLean AB, Nicol LA, Hodgins MB (1990) Immunohistochemical localization of estrogen receptors in the vulva and vagina. J Reprod Med 35(11):1015–1016

12. Raine-Fenning NJ, Brincat MP, Muscat-Baron Y (2003) Skin aging and menopause. Am J Clin Dermatol 4(6):371–378

13. Hasselquist MB, Goldberg N, Schroeter A, Spelsberg TC (1980) Isolation and characterization of estrogen receptor in human skin. J Clin Endocrinol Metab 50:76–82

14. Thornton MJ (2002) The biological actions of estrogens on skin. Exp Dermatol 11:487–502

15. Punnonen R, Lovgren T, Kouvonen I (1980) Demonstration of estrogen receptors in the skin. J Endocrinol Invest 3(3):217–221

16. Barbieri R, Ryan K (1999) The menstural cycle. In: Ryan K, Berkowitz R, Barbieri R, Dunaif A (eds) Kistner's gynecology and women's health, 7th edn. Mosby, St Louis, pp 23–30

17. Grumbach MM, Styne DM (2003) Puberty: ontogeny, neuroendocrinology, physiology, and disorders. In: Larsen PR, Kronenberg HM, Melmed S, Polonsky KS (eds) Williams textbook of endocrinology. Elsevier Science, Philadelphia, p 1143

18. Walsh BW, Ginsburg ES (1999) Menopause. In: Ryan K, Berkowitz R, Barbieri R, Dunaif A (eds) Kistner's gynecology and women's health, 7th edn. Mosby, St Louis, pp 540–542

19. Nelson LR, Bulun SE (2001) Estrogen production and action. J Am Acad Dermatol 45 [3 Suppl]:S116–S1124

20. Yen SCC (1977) The biology of menopause. J Reprod Med 18:287–296

21. Fletcher SW, Colditz G (2002) Failure of estrogen plus progestin therapy for prevention. JAMA 288:366–368

22. Rymer J, Wilson R, Ballard K (2003) Making decisions about hormone replacement therapy. BMJ 326(7384):322–326

23. Stampfer MJ, Colditz GA, Willett WC, et al (1991) Postmenopausal estrogen therapy and cardiovascular disease. Ten-year follow-up from the nurses' health study. N Engl J Med 325:756–762

24. Solomon CG, Dluhy RG (2003) Rethinking postmenopausal hormone therapy. N Engl J Med 348:579–580

25. Rossouw JE, Anderson GL, Prentice RL, et al (2002) Risk and benefits of estrogen plus progestin in healthy postmenopausal women: principal results from the women's health initiative randomized controlled trial. JAMA 288(3):321–333

26. Hulley S, Grady D, Bush T, et al (1998) Randomized trial of estrogen plus progestin for secondary prevention of coronary heart disease in postmenopausal women. Heart and Estrogen/progestin Replacement Study (HERS) Research Group. JAMA 280:605–613

27. Rivera-Woll LM, Davis SR (2004) Post menopausal hormone therapy: the pros and cons. Int Med J 34:109–114

28. Affinito P, Palomba S, Sorrentino C, et al (1999) Effects of postmenopausal hypoestrogenism on skin collagen. Maturitas 33:239–247

29. Brincat M, Moniz CJ, Studd JW, et al (1985) Long-term effects of the menopause and sex hormones on skin thickness. Br J Obstet Gynaecol 92(3):256–259

30. Brincat M, Kabalan S, Studd JW, et al (1987) A study of the decrease of skin collagen content, skin thickness, and bone mass in the postmenopausal woman. Obstet Gynecol 70:840–845

31. Maheux R, Naud F, Rioux M, et al (1994) A randomized, double-blind, placebo-controlled study on the effect of conjugated estrogens on skin thickness. Am J Obstet Gynecol 170(2):642–649

32. Castelo-Branco C, Duran M, Gonzalez-Merlo J (1992) Skin collagen changes related to age and hormone replacement therapy. Maturitas 15(2):113–119

33. Sauerbronn AVD, Fonseca AM, Bagnoli VR, et al (2000) The effects of systemic hormonal replacement therapy on the skin of postmenopausal women. Int J Gynecol Obstet 68:35–41

34. Chotnopparatpattara P, Panyakhamlerd K, Taechakraichana N, et al (2001) An effect of hormone replacement therapy on skin thickness in early postmenopausal women. J Med Assoc Thai 84(9):1275–1280

35. Varila E, Rantala I, Oikarinen A, et al (1995) The effect of topical oestradiol on skin collagen of postmenopausal women. Br J Obstet Gynaecol 102(12):985–989

36. Callens A, Vaillant L, Lecomte P, et al (1996) Does hormonal aging exist? A study of the influence of different hormone therapy regimens on the skin of postmentopausal women using non-invasive measurement techniques. Dermatology 193(4):289–294

37. Bolognia JL, Braverman IM, Rosseau ME, Sarrel PM (1989) Skin changes in menopause. Maturitas 11:295–304

38. Brincat M, Versi E, Moniz F, et al (1987) Skin collagen changes in postmenopausal women receiving different regimens of estrogen therapy. Obstet Gynecol 70:123–127

39. Brincat M, Versi E, O'Dowd T, et al (1987) Skin collagen changes in post-menopausal women receiving oestradiol gel. Maturitas 9(1):1–5

40. Brincat M, Yuen AW, Studd JW, et al (1987) Response of skin thickness and metacarpal index to estradiol therapy in postmenopausal women. Obstet Gynecol 70(4):538–541

41. Haapasaari KM, Raudaskoski T, Kallioinen M, et al (1997) Systemic therapy with estrogen or estro-

11

gen with progestin has no effect on skin collagen in postmenopausal women. Maturitas 27:153–162

42. Oikarinen A (2000) Systemic estrogens have no conclusive beneficial effect on human skin connective tissue. Acta Obstet Gynecol Scand 79:250–254

43. Dunn L, Damesyn M, Moore A, et al (1997) Does estrogen prevent skin aging? Results from the First National and Health Nutritional Examination Survey. Arch Dermatol 133:339–342

44. Schmidt JB, Binder M, Demschik G, et al (1996) Treatment of skin aging with topical estrogens. Int J Dermatol 35:669–674

45. Pierard-Franchimont C, Letawe C, Goffin V, et al (1995) Skin water-holding capacity and transdermal estrogen therapy for menopause: a pilot study. Maturitas 22(2):151–154

46. Anttinen H, Orava S, Ryhanen L, et al (1973) Assay of protocollagen lysyl hydroxylase activity in the skin of human subjects and changes in the activity with age. Clin Chim Acta 47(2):289–294

47. Shah MG, Maibach HI (2001) Estrogen and skin: an overview. Am J Clin Dermatol 2(3):143–150

48. Grosman N, Hvidberg E, Schou J (1971) The effect of oestrogenic treatment on the acid mucopolysaccharide pattern in skin of mice. Acta Pharmacol Toxicol 30(5):458–464

49. Danforth DN, Veis A, Breen M, et al (1974) The effect of pregnancy and labor on the human cervix: changes in collagen, glycoproteins, and glycosaminoglycans. Am J Obstet Gynecol 120(5):641–651

50. Denda M, Koyama J, Hori J, et al (1993) Age- and sex-dependent change in stratum corneum sphingolipids. Arch Dermatol Res 285(7):415–417

51. Sator PG, Schmidt JB, Sator MO, et al (2001) The influence of hormone replacement therapy on skin ageing: a pilot study. Maturitas 39(1):43–55

52. Henry F, Pierard-Franchimont C, Cauwenbergh G, et al (1997) Age-related changes in facial skin contours and rheology. J Am Geriatr Soc 45(2):220–222

53. Pierard GE, Letawe C, Dowlati A, et al (1995) Effect of hormone replacement therapy for menopause on the mechanical properties of skin. J Am Geriatr Soc 43(6):662–665

54. Creidi P, Faivre B, Agache P, et al (1994) Effect of a conjugated oestrogen (Premarin) cream on ageing facial skin. A comparative study with a placebo cream. Maturitas 19(3):211–223

55. Punnonen R, Vaajalahti P, Teisala K (1987) Local oestriol treatment improves the structure of elastic fibers in the skin of postmenopausal women. Ann Chir Gynaecol Suppl 202:39–41

56. Ashcroft GS, Lei K, Jin W, et al (2000) Secretory leukocyte protease inhibitor mediates non-redundant functions necessary for normal wound healing. Nat Med 6(10):1147–1153

57. Ashcroft GS, Dodsworth J, Boxtel E, et al (1997) Estrogen accelerates cutaneous wound healing associated with an increase n TGF-β1 levels. Nat Med 3:1209–1215

58. Herrick SE, Ashcroft GS, Ireland G, et al (1996) Upregulation of elastase in acute wounds of healthy aged humans and chronic venous leg ulcers is associated with matrix degradation. Lab Invest 77:281–288

59. Ashcroft GS, Horan MA, Ferguson MW, et al (1997) Aging is associated with reduced deposition of specific extracellular matrix components, an upregulation of angiogenesis, and an altered inflammatory response in a murine incisional wound healing model. J Invest Dermatol 108:430–437

60. Jorgensen O, Schmidt A (1962) Influence of sex hormones on granulation tissue formation and on healing of linear wounds. Acta Chir Scand 124:1–10

61. Calvin M, Dyson M, Rymer J, et al (1998) The effects of ovarian hormone deficiency on wound contraction in a rat model. Br J Obstet Gynaecol 105:223–227

62. Ashcroft GS, Greenwell-Wild T, Horan MA, et al (1999) Topical estrogen accelerates cutaneous wound healing in aged humans associated with an altered inflammatory response. Am J Pathol 155(4):1137–1146

63. Margolis DJ, Knauss J, Bilker W (2002) Hormone replacement therapy and prevention of pressure ulcers and venous leg ulcers. Lancet 23:675–677

64. Berard A, Kahn SR, Abenhaim L (2001) Is hormone replacement therapy protective for venous ulcer of the lower limbs? Pharmacoepidemiol Drug Saf 10:245–251

65. Ashcroft GS, Mills SJ, Lei KJ, et al (2003) Estrogen modulates cutaneous wound healing by downregulating macrophage migration inhibitory factor. J Clin Invest 111:1309–1318

66. Shah M, Foreman DM, Ferguson MWJ (1992) Control of scarring in adult wounds by neutralizing antibody to transforming growth factor β. Lancet 339:213–214

67. MacLennan AH, Taylor AW, Wilson DH (1995) Changes in the use of hormone replacement therapy in South Australia. Med J Aust 162:420–422

68. Bachman GA (1997) A new option for managing urogenital atrophy in postmenopausal women. Cont Obstet Gynecol 42:13–28

69. Writing Group for the Womens Health Initiative Investigators (2002) Risks and benefits of estrogen plus progestin in healthy postmenopausal women. JAMA 288:321–333

70. Viscoli CM, Brass LM, Kernan WW, et al (2001) Clinical trial of estrogen replacement after ischemic stroke. N Engl J Med 345:1243–1249

Chapter 12

Skin Aging in Three Dimensions

12

Niels C. Krejci-Papa, Robert C. Langdon

Contents

12.1 Introduction

Patients and physicians have a natural inclination to concentrate on epidermal and dermal features when evaluating intrinsic aging and photoaging of the skin. However, the cutaneous fat and the cutaneous muscles with their aponeuroses (superficial musculoaponeurotic system, SMAS) have an enormous impact on apparent age, and can be manipulated by dermasurgeons and other cosmetic surgeons with great benefit. Volume and contour changes are as important or more important than two-dimensional surface characteristics when perceiving a person's age (Fig. 12.1). In this chapter three-dimensional changes in the context of overall skin aging are evaluated, and current treatment approaches are described.

12.2 Age-Related Changes at Various Depths

The skin of the face and neck differs from skin at most other body sites in that it has retained the cutaneous musculature that is found as a the "panniculus carnosus" in the hide of lower mammalian ancestors [7]. Cutaneous muscles are a phylogenetic and ontogenetic part of the skin and must be differentiated from skeletal muscles whose function is to move bones (Fig. 12.2). In humans, the facial muscles have evolved beyond the functional opening and closing of eyes, mouth and nose to projecting emotion for social interaction.

A simplified anatomical organization of the face by layers thus includes: the epidermis, dermis, the cutaneous fat, the cutaneous muscles, which are collectively termed the SMAS, the deep skeletal muscles of mastication, the perios-

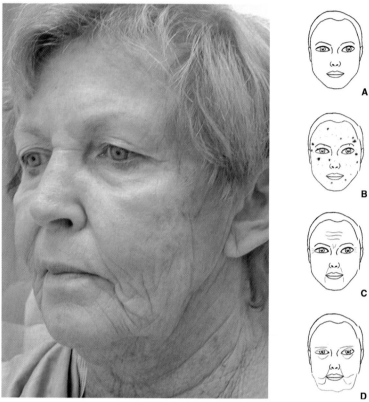

12

Fig. 12.1. Perceived age is affected by changes in the epidermis, dermis and facial contour "the third dimension". This patient desires to focus treatment on those features that contribute most to perceived age. *A* Youthful face. *B* Epidermal changes only: mottled pigmentation, seborrheic keratoses. Perceived age is only minimally impacted. *C* Dermal changes only: ingrained static and motion-induced rhytids. Perceived age is considerably impacted. *D* Contour changes only: brow ptosis, upper and lower lid festooning, ptosis of the nasal tip and sagging of the cheek and jowls. Perceived age is dramatically impacted

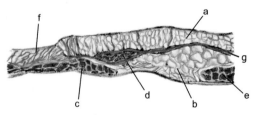

Fig. 12.2. Horizontal cross section through facial skin at the level of the right upper lip and cheek: *a* subcutaneous fat overlying the cutaneous musculature (SMAS); *b* deep "skeletal" fat under the cutaneous musculature; *c* orbicularis oris muscle; *d* elevator labii muscle; *e* masseter muscle (a skeletal muscle); *f* fibrous septae which anchor the dermis to the SMAS and convert muscle contractions to dermal wrinkling or folding; *g* cutaneous musculature (*SMAS* superficial musculoaponeurotic system)

teum, and bone. Each of these components undergoes a unique set of age-related changes.

12.2.1 Epidermal Changes

Changes in the epidermis appear as irregularities in the surface texture and skin color (Fig. 12.3).

The interappendageal epidermis thins over time, and flattened rete ridges produce a fine crinkling. An irregular stratum corneum with slight scaling leads to increased scattering of light and thus the projection of a dull, dry appearance. Papules and plaques from seborrheic

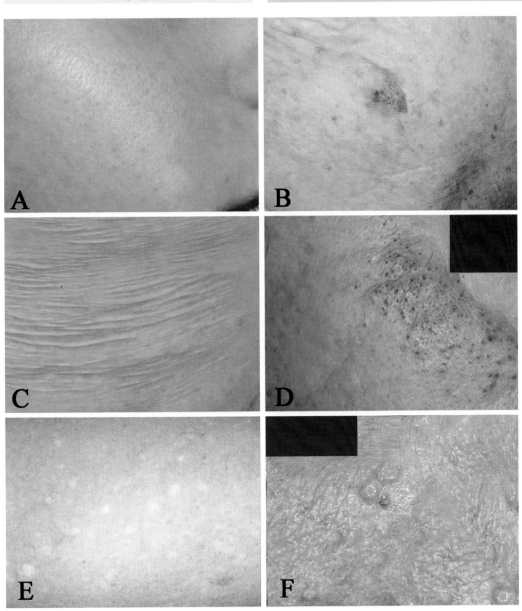

Fig. 12.3. Epidermal changes associated with intrinsic aging and photo-aging: **A** youthful skin; **B** mottled pigmentation and solar lentigo; **C** atrophy of epidermis and dermis resulting in "cigarette-paper" wrinkling; **D** actinic comedones; **E** guttate hypomelanosis; **F** sebaceous hyperplasia (photographs **D**, **E**, and **F** by Thomas Habif, MD, www.dermnet.org)

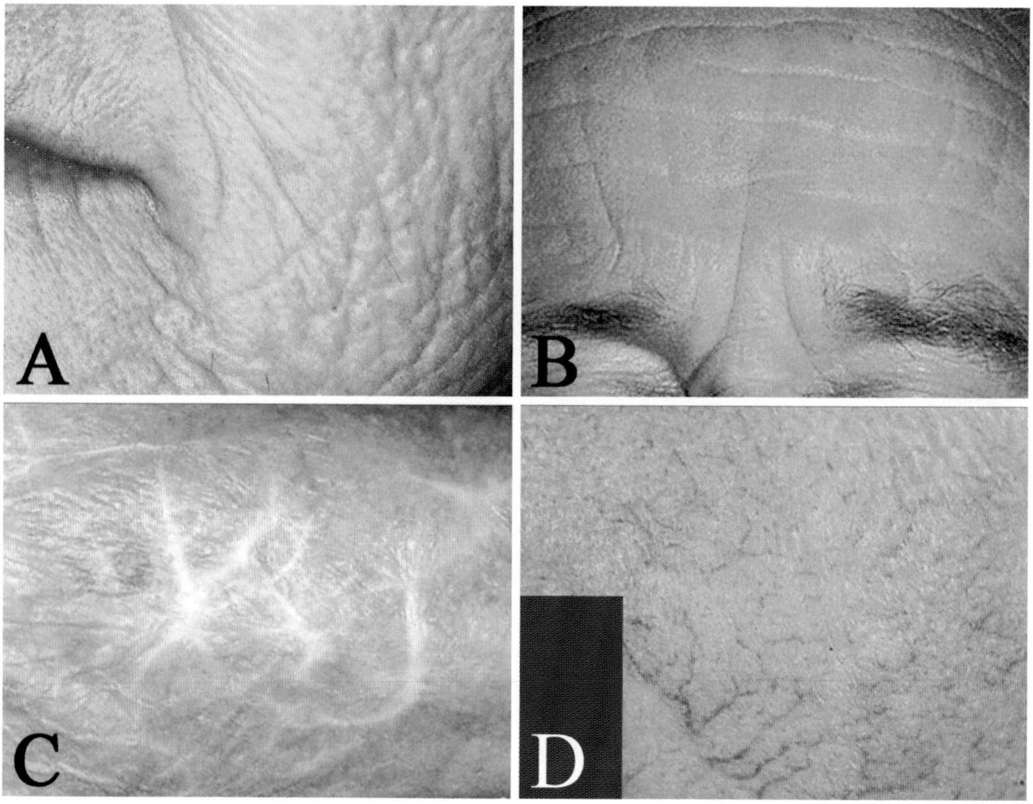

Fig. 12.4 Dermal changes associated with intrinsic aging and photo aging: **A** ingrained perpendicular rhytids in lax skin; **B** ingrained motion-induced rhytids (horizontal) perpendicular to muscle action and sleep lines (vertical); **C** pseudoscars; **D** superficial telangiectasias (photographs **A–D** by Thomas Habif, MD, www.dermnet.org)

("senile") keratoses and actinic keratoses are hallmarks of aged skin.

12.2.2 Dermal Changes

Age-related changes in the dermis include the loss of elasticity which, in conjunction with underlying muscle movement, produces the ingrained "motion-induced" rhytids such as horizontal forehead lines (Fig. 12.4). Similarly, repeated folding of skin by sleeping in the same position on the side of the face ingrains "sleep lines" [5].

12.2.3 Fat Changes

Anatomical layers deeper than the superficial dermis are not reached by ultraviolet light and thus undergo strictly "intrinsic" aging as opposed to "photoaging".

Fat compartments of various facial regions age differently. Fat atrophy occurs in the periorbital, forehead, buccal, temporal and perioral areas. Fat accumulation is seen submentally, in the jowls, lateral nasolabial folds, lateral labiomental creases and lateral malar areas (Fig. 12.5). Some areas, such as the infraorbital area, may present with either atrophy or hypertrophy/festooning [3].

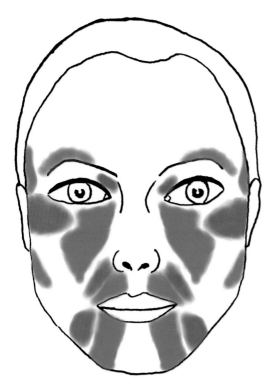

Fig. 12.5. Fat compartments of the face tend to either atrophy with age (*green*) or show fat accumulation (*purple*). Thus "hollow cheeks" may be adjacent to adipose jowls (used with permission from Ref. 3)

Other fat compartments are more noticeable for their movement than for their relative atrophy or hypertrophy. The malar fat pad is a triangular body of cheek fat with the base at the nasolabial fold and the apex directed towards the malar eminence. The fat pad is firmly adherent to the cheek skin, but with age its connection to the SMAS is weakened, presumably due to the shearing forces generated by the repeated contraction of the lip levators [6]. As the malar fat pad descends with age, it produces a concavity at the infraorbital region and a deep bulging as it presses against the fixed restraint of the nasolabial creases.

12.2.4 SMAS Changes

The SMAS is a discrete fibromuscular layer (Fig. 12.2, g) enveloping and interlinking the muscles of facial expression [12]. There is controversy over how much the muscular component of the SMAS elongates while it thins with age. However, both the aponeurotic parts and the fibrous trabeculae that attach the dermis (see Fig. 12.2, f) to the SMAS stretch over time, resulting in sagging of the overlying skin (Fig. 12.6).

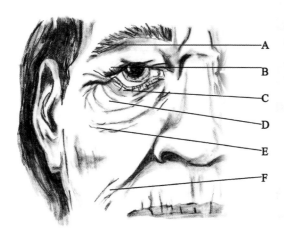

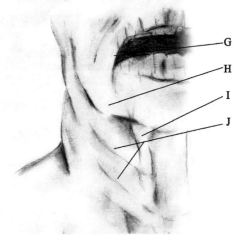

Fig. 12.6. The superficial musculoaponeurotic system (SMAS) is the compartment of the skin that is mainly responsible for its anchorage and suspension on the bony skeleton of the scull. Three-dimensional/contour changes related to gravitational descent include: *A* brow ptosis, *B* hooding of the upper eyelids, *C* ectropion, *D* bulging of the lower eyelid, *E* festooning, *F* heavy or overhanging nasolabial fold, *G* ptosis of angle of the mouth, *H* jowls, *I* "turkey gobbler" hanging skin-fold, *J* horizontal neck folds

12

The zygomatic bones have long horizontal attachments to the SMAS-retaining ligaments, which effectively separate the gravitational descent of the upper and mid/lower face.

Unlike the upper face and mid-face, the SMAS of the lower face and neck (much of it platysma) does not overlie periosteum, and the predominant plane of tissue sagging occurs between the SMAS and the underlying deep cervicofacial fascia. Here the musculature itself shows conversion of elastic to fibrous connective tissue, and as it stretches, it takes the overlying fat and dermis down as a unit [10].

12.2.5 Bone and Cartilage Changes

In post-adulthood, total facial height increases with age when measured radiographically [1]. Other age-related changes include maxillary bone atrophy, reduction of chin projection, reduction of the mandibular angle, and loss of the inferior orbital rim projection. The maxillary and mandibular changes are most pronounced in patients with edentulism [11]. Cartilaginous changes contribute little to aged appearance, with the possible exception of the nasal septum and its role in the shortening of the columella and the associated ptosis of the nasal tip.

12.3 Age-Related Changes of Anthropometric Landmarks

Evaluation of volume and contour changes is based on the recording of both qualitative and quantitative facial features. Anthroscopy refers to the recording of qualitative signs such as facial shape, symmetry, and convexity of the nasal dorsum. Anthropometry refers to the quantitative measurement of distances, angles, proportions and volumes [4]. Because volume loss and sagging of an anatomical area are most evident at its inferior border, the resultant changes are often discussed referring to the inferior landmarks: e.g. forehead descent as brow ptosis, or cheek descent as jowl ptosis or "jowling".

Modern cosmetic surgery has attempted to quantify the benefit of its procedures, but, for the most part, has been unable to capture a "rejuvenation factor" and still relies heavily on before-and-after photographs.

12.4 Concepts in Three-Dimensional Contouring

While there are some areas of fat accumulation that will benefit from a volume reduction, it is flattening, volume loss and sagging skin that are the hallmarks of facial aging. The three predominant approaches for correction include: volume restoration, pulling and tightening of sagging skin, and skin tightening via thermal contraction of collagen.

12.4.1 Volume Restoration

Re-creation of facial convexities that reflect light and that reduce the shadows of facial concavities has a strong effect on the perceived age of a person [3]. Volume restoration is mostly performed using injectable fillers, fat transplants, fat flaps, or artificial implants. Because more tissue is needed to drape the increased volume, wrinkles diminish, folds are reduced and inferior landmarks are lifted.

12.4.2 Pulling and Tightening of Sagging Skin

This approach, which is best exemplified by the traditional face lift, employs a combination of fixation, plication, or imbrication of the approximated layers of the SMAS to reposition skin and associated landmarks. These surgeries produce excellent results as long as care is taken not to accentuate atrophic changes. If plication is used instead of resection, a filling effect can be achieved that reduces the flattening effect of the stretch.

12.4.3 Skin Tightening via Thermal Contraction of Collagen

This rejuvenation technique is an important benefit of CO_2 laser resurfacing. In 2004, the FDA approved the radiofrequency heating of the dermis with simultaneous epidermal cooling. This avoids the prolonged healing that is required for reepithelialization.

12.5 Contouring via Volume Replacement

12.5.1 Volume Replacement using Injectable Fillers

The number of injectable fillers is legion and ever-growing. FDA-approved and non-approved materials are available in many viscosities that allow treatment from the filling of fine upper lip lines to the massive volume replacement needed to correct HIV lipodystrophy. The longevity of the effect depends on the chemical make-up of the filler and ranges from a few weeks to permanence. Excellent reviews on this topic have been published [8, 13].

12.5.2 Volume Replacement with Autologous Fat

Autologous fat is often the material of choice for facial volume augmentation because of its availability, relatively low cost/volume ratio, and natural feel. Fat sheets may be used as either free flaps [15] or local flaps [11]. Most often, however, fat is harvested from a remote donor site via cannula aspiration and implanted via injection of suspended fat tissue. All layers from the periosteum to the dermis can receive injected fat but it survives best adjacent to the facial musculature [3]. Some surgeons strive to thread small fat pockets into the individual muscles – a method developed by R.E. Amar [2] and known as FAMI (fat autograft muscle injection). While injectable synthetic fillers are mostly used to address a single area of concern, the trend in fat grafting is to restore the complete face to a fuller, more youthful contour.

12.5.3 Volume Replacement with Solid Implants

Solid implants are inserted either by a trochar or via skin incisions. Implants are often manufactured from inert synthetics and produce a permanent volume increase. Two important examples are the soft and pliable expanded polytetrafluoroethylene (ePTFE) tubes, and the host of firm, anatomically contoured silicone implants for accentuation of natural facial convexities.

ePTFE was initially developed as Gore-Tex and is now mainly employed as a porous tube, which can be placed subdermally with a trochar. It allows ingrowth of fibroblasts and thus integrates into the connective tissue matrix [9]. The main indications are upper and lower lip augmentation and treatment of nasolabial folds

Silicone is the umbrella name given to a large group of compounds based on elemental silicon. Medical silicones are based on dimethylsiloxane chains which can be crosslinked to any viscosity, from watery to gel to rubbery to hard. Most alloplastic silicone implants in cosmetic surgery are opaque, non-porous, rubbery designs. They are manufactured in various anatomical shapes and sizes, but are also easily carved using a scalpel. Solid silicone implants are often sutured in place and, over time, induce the formation of a surrounding fibrous capsule [14].

12.6 Contouring via Lifting Techniques

12.6.1 The Brow or Forehead Lift

The attachment of the SMAS to the zygomatic bones separates the age-related descent of the upper face and the mid/lower face – resulting in the traditional separation of forehead lift and face lift. Descent of the forehead tissues is most notable through the drop in brow position that projects a tired and sad expression. All patients presenting for upper eyelid blepharoplasty

12

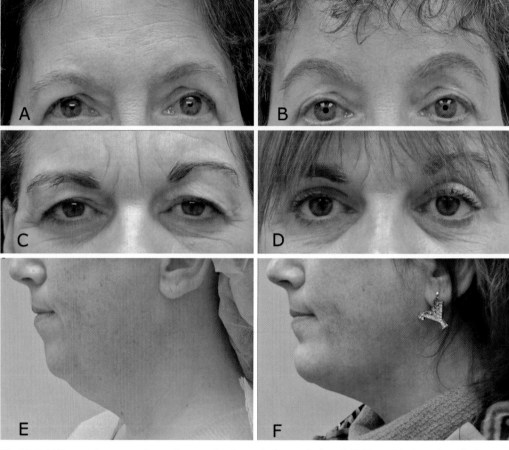

Fig. 12.7. Lifting and resuspension, volume reduction and volume addition are employed to restore a youthful appearance in all three dimensions. **A, B** Brow lift before and after. **C, D** Upper blepharoplasty before and after. **E, F** Neck liposuction and chin implant before and after

should be evaluated for brow position, as it is often a major factor in upper lid fullness, but will not be corrected by a blepharoplasty alone.

The commonly used designations for brow lifting techniques refer to the location of the initial incision. Examples are: the direct brow lift (the incision is directly above the brow), and the transpalpebral, temporal, mid-forehead and coronal brow lifts. There are variations in the level of dissection (subcutaneous, subgaleal, subperiosteal), the inclusion of a subperiosteal release of the arcus marginalis and the optional removal/destruction of the medial brow depressor (frown) muscles. There are also variations in the method of elevation: excision of skin alone

or excision of skin and frontalis muscle. Both techniques can be combined with a fixation to the periosteum.

Recently, open approaches have lost popularity and are being replaced with endoscopic brow lifts, in which five small ports are hidden in the hair line. Most of the options listed above apply also to the endoscopic surgeries (Fig. 12.7A, B).

12.6.2 Upper Eyelid Blepharoplasty

Upper lid fullness and ptosis that is independent of brow ptosis can be corrected by excision of excess palpebral skin (with or without orbicularis muscle), the reduction of the two retroseptal fat pads, and/or resuspension of the lacrimal gland. The incision is commonly placed in the eyelid crease where it heals with an imperceptible scar (Fig. 12.7C, D)

12.6.3 Lower Eyelid Blepharoplasty

Aging produces a descent of the lateral canthal ligaments, which leads to laxity of the lower lid margin with possible ectropion and sublimbal scleral show. The frequently seen bulging of the lower lid results from a combination of excess/stretched skin, palpebral edema and fat pad pseudoherniation through a weakened orbital septum. Soft tissue atrophy in the maxillary area inferior to the lower lid can also contribute to the perception of prominence of lower lid bulging. Corrective procedures include: lateral canthopexy (usually performed through an incision at the upper lateral lid), reduction or repositioning of the fat pads (usually through a transconjunctival approach), and skin/orbicularis resection (usually through a subciliary incision).

12.6.4 Face/Neck Lift

Lifting procedures of the cheeks and neck are often combined and intimately affect each other. Most incisions start in the temporal hairline and proceed preauricularly, perilobularly and retroauricularly. The commonly used designations for face lifts refer to the level of mobilization. Skin-only face lifts include less-extensive dissection and no plication of the SMAS. They do not produce long-lasting results but are occasionally performed as touch-ups after a standard lift. The traditional or "SMAS" face lift uses a subcutaneous dissection at the surface of the cutaneous muscles. The SMAS is then shortened using imbrication or plication or a purse string suture, which can be combined with a periosteal fixation. The skin is then re-draped and excess is removed at the incision line. "Deep plane" facelifts imply an additional sub-SMAS mobilization over the parotid gland.

A subperiosteal dissection can be done using either open or endoscopic techniques. The goal is to release the SMAS adhesions to the zygomatic arch and to lift the mid-lower face in conjunction with a forehead lift. This adds an additional vertical vector to the face lift that is difficult to achieve otherwise (Fig. 12.7E, F).

12.7 Contouring via Thermal Skin Contraction

When quickly heated to approximately 60°C, cutaneous collagen type I shrinks and does not re-nature spontaneously. Instead, the shrunken collagen induces dermal remodeling that stabilizes the effect. This contraction is dramatically visible during scanned CO_2 laser resurfacing. When adjacent areas of skin squares are sequentially exposed to the laser, there is visible shrinking during the short laser impact.

Many attempts have been made to employ this dermal effect without affecting the epidermis and the concept is frequently referred to as non-ablative tissue remodeling. The FDA-approved Thermage unit utilizes radiofrequency for dermal heating combined with epidermal cooling via a cold metal sheet. The resulting lift is often small, but there is no incisional scar or downtime for the patient. Because only the skin is contracted, the longevity of the tightening should resemble that of a resurfacing procedure, not that of a SMAS face lift. On the other hand, no contraindications to repeated or even frequent "additive" or "upkeep" treatments are known.

12.8 Summary

Patients and physicians are often inclined to view skin aging as a mostly dermal/epidermal process that is primarily the result of exposure to ultraviolet light. In the face, however, the distribution of subcutaneous fat and the unique cutaneous musculature create a third dimension that strongly influences the patient's apparent age.

References

1. Behrents R (ed) (1985) Growth in the aging cranio-facial skeleton. Craniofacial growth series. Center for Human Growth and Development, Michigan
2. Butterwick K, Lack E (2003) Facial volume restoration with the fat autograft muscle injection technique. Dermatol Surg 29:1019–1026
3. Donofrio L (2000) Fat distribution. Dermatol Surg 26:1107–1112
4. Farkas L (1994) Anthropometry of the head and face. Raven Press, New York
5. Fulton J (1999) Sleep lines. Dermatol Surg 25:59–62
6. Gosain A (1996) A dynamic analysis of changes in the nasolabial fold using magnetic resonance imaging: implications for facial rejuvenation and facial animation surgery. Plast Reconstr Surg 98:622–636
7. Haeckel E (2004) The evolution of man. IndyPublish, McLean
8. Jordan D (2003) Soft-tissue fillers for wrinkles, folds and volume augmentation. Can J Ophthalmol 38:285–288
9. Miller P (2000) Softform for facial rejuvenation: historical review, operative techniques, and recent advances. Facial Plast Surg 16:23–28
10. Owsley J, Fiala T (1997) Update: lifting the malar fat pad for correction of prominent nasolabial folds. Plast Reconstr Surg 100:715–722
11. Ramirez OM (2001) Full face rejuvenation in three dimensions: a "face-lifting" for the new millennium. Aesthetic Plast Surg 25:152–164
12. Salashe SJ, Bernstein G, Senkarik M (1988) Surgical anatomy of the skin. Appleton & Lange, Norwalk
13. Sclafani AP, Romo T 3rd (2000) Injectable fillers for facial soft tissue enhancement. Facial Plast Surg 16:29–34
14. Terino E, Flowers R (2000) The art of alloplastic facial contouring, Mosby, St Louis
15. Williams H (1979) Free dermal fat flaps to the face. Ann Plast Surg. 3:1–12

12

Chapter 13

Retinoid Therapy for Photoaging

Laure Rittié, Gary J. Fisher, John J. Voorhees

13

Contents

13.1 Introduction

Since the initial report that described that topical all-*trans*-retinoic acid (RA) had a beneficial effect on photoaging [1], the effectiveness of topical all-*trans*-RA in the treatment of photodamaged skin has been demonstrated in a large number of controlled clinical studies [2–7]. Discoveries during the past 10 years have provided detailed knowledge regarding the fundamental molecular mechanisms of action of all-*trans*-RA in mammalian tissues. In spite of this knowledge, large gaps exist in our understanding of the molecular basis by which all-*trans*-RA improves the appearance of photodamaged skin.

The term photoaging relates to alterations that occur in skin that is chronically exposed to sunlight. In sun-exposed skin, photoaging is superimposed on natural, intrinsic aging, which occurs in all skin. Subject-matched clinical, histological, and biochemical analysis of sun-exposed and sun-protected skin readily reveals that sun exposure undermines the appearance and function of human skin to a much greater degree than natural aging. Photoaged skin appears dry, wrinkled, lax and unevenly pigmented, and has brown spots [8]. Histologically, photoaged skin shows variable epidermal thickness, large and irregular grouped melanocytes, lack of compaction of the dermis, elastosis, and mild inflammatory infiltrate. Photoaging is a cumulative process, and as such, its severity is highly correlated with age. Conversely, photoaging is inversely correlated with skin pigmentation, since darker skin provides a more effective barrier against penetration of sunlight. During the past 20 years, great progress has been made in our understanding of the molecular mechanisms responsible for photoaging.

In this chapter, we first describe the molecular mechanisms by which ultraviolet (UV) irradiation from the sun damages human skin, focusing on damage to collagens, which are the major structural components of skin connective tissue. We do not describe in detail UV damage to DNA, which is addressed in chapter 4. We present evidence that topically applied retinoids can both prevent UV-induced skin damage and repair photodamaged skin. In addition, we describe our current understanding of the mechanism of action of retinoids. Finally, we review some of the side effects associated with retinoid therapy.

13.2 UV-Mediated Damage to Skin Connective Tissue

UV irradiation from the sun causes a large number of deleterious effects to human skin, including sunburn, immunosuppression, and skin cancer, in addition to premature aging. Solar UV radiation that reaches the surface of the earth is subdivided by wavelength into UVB (290–320 nm), UVA2 (320–340 nm) and UVA1 (340–400 nm). The energy inherent in UV radiation can be absorbed by many different cellular components within the skin, and consequently elicits a variety of cellular responses brought about by coordinated activation of signal transduction pathways that result in alterations in gene expression. A common initiating event following UV irradiation is the photochemical generation of reactive oxygen species (ROS), which can modify different cellular components (i.e. DNA, proteins, lipids), and cause oxidative stress. In this section, we review the mechanisms by which UV light activates specific signaling pathways, and thereby modifies gene transcription and damages human skin connective tissue.

13.2.1 UV Irradiation Induces Activation of Signaling Pathways

The primary mechanism by which UV radiation initiates molecular responses in human skin is via photochemical generation of ROS, including superoxide anion (O_2^-), hydrogen peroxide (H_2O_2), hydroxyl radical (HO·), and singlet oxygen ($^1O_2$) [9]. ROS are produced by energy and/or electron transfer from UV-absorbing chromophores in the skin. Many chromophores have been proposed as endogenous ROS generators, including tryptophan [10], nicotinamide adenine dinucleotide phosphate (NADPH), riboflavin, porphyrin [11], and trans-urocanic acid [12]. ROS may also be generated by UV irradiation of dermal collagen and elastin [13], and soluble proteins modified by advanced glycation end products (AGE), such as albumin [14, 15]. Increased ROS production is detected within one minute after UVA irradiation of mouse skin in vivo, and can be prevented by topical application of superoxide dismutase [16].

In human skin, UV-induced liberation of ROS is responsible for stimulation of numerous signal transduction pathways via activation of cell surface cytokine and growth factor receptors. These include epidermal growth factor receptor (EGF-R) [17, 18], insulin receptor [19], platelet derived growth factor receptor (PDGF-R) [20], tumor necrosis factor-α receptor (TNF-α-R) and interleukin-1 receptor (IL-1R) [21]. UV irradiation of skin cells induces a substantial response due to simultaneous activation of multiple receptors [21]. EGF-R activation occurs within 15 minutes following UV exposure (twice the minimal erythema dose or MED) of human skin in vivo and remains elevated for two to four hours [22].

UV-mediated activation of EGF-R is ligand-independent, since the ligand-binding domain is not required for UV activation of the receptor [20]. However, direct activation of EGF-R by UV light has never been demonstrated. Alternatively, it has been suggested that UV-induced growth factor and cytokine receptor activation is mediated by ROS-mediated inactivation of protein tyrosine phosphatases (PTPs) [20, 23]. PTPs act on tyrosine kinase receptors to maintain a low basal phosphorylation state of the receptor. PTPs act in concert with tyrosine kinases to maintain a dynamic equilibrium between phosphorylation and dephosphorylation of receptors. ROS formed after UV irradiation inactivate PTPs via reversible oxidation of a critical cysteine residue within the catalytic site, common to all PTPs [24, 25]. Thus, UV irradiation of cells can increase the half-life of EGF-R phosphorylated tyrosine(s) [20].

Recently, it has been shown that in addition to reversible oxidative inactivation of PTPs, UVA and UVB can trigger degradation of certain PTPs in cultured cells. This degradation is mediated by the protease calpain, and requires both activation of calpain and the oxidative alteration of PTP by UV irradiation [26]. The PTPs involved in responses of human skin to UV irradiation remain to be identified.

Downstream signaling induced by UV-mediated activation of cell surface receptors is similar to that which follows ligand binding. Autophosphorylation of cell surface receptors results in recruitment of adaptor and regulatory proteins that mediate downstream signaling, including activation of the three families of mitogen-activated protein kinases (MAPKs): extracellular signal-regulated kinase (ERK), c-Jun amino-terminal kinase (JNK), and p38. Activation of protein kinase-mediated signaling pathways occurs within one hour after UV irradiation of human skin in vivo, and is maximal four hours after UV irradiation. Activation of all three MAP kinase modules occurs throughout the epidermis and upper dermis [27]. Activated JNK and p38 catalyze the functional activation of transcription factors c-Jun, and activating transcription factor-2 (ATF-2), respectively. Activated ATF-2 binds c-Jun promoter and upregulates c-Jun protein transcription [28]. C-Jun partners with c-Fos, which is constitutively expressed in human skin and is not further elevated after UV irradiation [22], to form activator protein (AP)-1 transcription factor complex. Stimulation of AP-1 by UV irradiation induces gene expression of AP-1-regulated matrix metalloproteinases (MMPs), i.e. MMP-1 (interstitial collagenase), MMP-3 (stromelysin 1), and MMP-9 (gelatinase B) [29]. In parallel, AP-1 negatively regulates type I procollagen synthesis by dermal fibroblasts [30]. MMP induction occurs throughout the epidermis and upper dermis. Induction of MMP-1, -3, and -9 mRNA is detected within eight hours following UV irradiation of human skin in vivo, and is followed by increased protein synthesis, and enzyme ac-tivities [29, 31]. MMP levels remain elevated for two to three days after a single acute UV exposure in human skin in vivo [29]. Increased levels of endogenous glutathione induced by topical application of N-acetyl cysteine (NAC) reduce UV-induced activation of c-Jun and consequent induction of MMPs [32]. This observation is consistent with the involvement of ROS in UV-mediated activation of MMPs.

UV irradiation of skin also upregulates nuclear factor-κB (NF-κB). The NF-κB pathway regulates expression of genes primarily involved in immune and inflammatory responses. In the absence of stimuli, NF-κB is sequestered in the cytoplasm, bound to its inhibitor IκB. Cytokine activation of NF-κB is mediated by stimulation of IκB kinase (IKK), which phosphorylates IκB. Phosphorylated IκB is rapidly degraded and releases NF-κB, which translocates into the nucleus where it activates target genes. However, this mechanism is not applicable to UV-induced NF-κB activation since no IKK activation is detected after UV irradiation [33, 34]. As opposed to TNF-α, UV irradiation activates NF-κB in a delayed (three to six hours after irradiation) and relatively weak manner [33, 34]. Recently, it has been demonstrated that UV irradiation downregulates IκB through inhibition of IκB synthesis [35]. In the absence of ongoing IκB synthesis, the cellular IκB content is slowly depleted by natural degradation, eventually leading to NF-κB activation. It has also been demonstrated that the carboxyterminal zinc finger domain of NEMO (regulatory subunit of IKK) is required for the UV-induced NF-κB signaling pathway [36]. Recently, it has been reported that the protein kinase MAPK-activated kinase-2, which is activated by p38, phosphorylates heat shock protein 27 (Hsp27), which in turn binds to IKKβ (one of the two catalytic subunits of IKK). Binding of Hsp27 to IKKβ inhibits IKK activity [37]. UV irradiation strongly induces p38 in human skin in vivo [28]. Therefore, it is possible that UV-induced p38 activation weakens NK-κB activation and downregulates IKK activity to below the level of detection.

Once translocated into the nucleus, NF-κB upregulates the synthesis of numerous cytokines including IL-1β, TNF-α, IL-6, and IL-8 [38]. These cytokines are secreted, and act in an autocrine and/or paracrine manner by binding to their cell surface receptors on epidermal and dermal cells, thereby activating AP-1 and NF-κB and amplifying the UV response.

13.2.2 UV Irradiation Induces Degradation of Skin Collagen

UV-induced activation of MMPs results in excessive degradation of extracellular matrix (ECM) components in the dermis. ECM is the substrate for cell adhesion, growth, migration, and differentiation, and provides mechanical support for tissues and organs. ECM proteins are secreted by fibroblasts in the dermis, and feedback mechanisms allow the fibroblasts to adapt to alterations in the ECM. This bidirectional interaction between ECM and ECM-supporting fibroblasts is critical for optimal function of skin connective tissue.

Type I collagen is the most abundant structural protein in skin connective tissue. It accounts for about 85% of total dermal protein. Type III collagen interacts with type I collagen, and is present at approximately one-tenth the level of type I collagen. Type I and III collagens are composed of three polypeptide chains, organized in a rod-like triple helical structure. Type I and type III collagens are synthesized by fibroblasts in the dermis, as precursor proteins (procollagens) containing globular domains at each of their ends. These globular portions, referred to as N-terminal and C-terminal propeptides, allow procollagens to be secreted from the cell as a soluble monomeric protein. Once secreted, the propeptides are enzymatically removed by specific proteases. Cleavage of the C-terminal propeptide precedes cleavage of the N-terminal propeptide. Removal of the propeptides allows the collagen molecules to spontaneously assemble into fibers. Newly formed collagen molecules assemble into existing collagen fibers, and undergo intermolecular crosslinking to form mature collagen fibers.

Collagen crosslinking is catalyzed by the enzyme lysyl oxidase [39]. Collagen fibers are arranged in orderly arrays in association with other ECM proteins to form collagen bundles.

Collagen bundles provide strength and resiliency to the skin, and are very resistant to enzymatic degradation. The half-life of skin collagen is estimated to be several years. Degradation of collagen fibers requires the action of collagenases. Three mammalian collagenases exist, MMP-1, MMP-8, and MMP-13, which have the unique capacity to initiate cleavage of mature collagen. As described above, UV irradiation induces MMP-1. MMP-8 is primarily expressed by neutrophils, which infiltrate skin in response to immune stimuli induced by UV irradiation. Although MMP-8 levels increase in human skin following UV irradiation as a consequence of neutrophil influx, MMP-8 remains largely in an inactive state, and therefore does not significantly contribute to the collagen degradation observed following UV exposure [27]. MMP-13 is not induced by UV irradiation in human skin in vivo. Therefore, MMP-1 is the primary collagenase that is responsible for initiation of collagen breakdown in UV-irradiated human skin. Once cleaved by a collagenase, fragmented collagen can be further degraded by elevated levels of MMP-3 and MMP-9 [40].

13.2.3 UV Irradiation Inhibits Production of New Collagen

In addition to causing degradation of mature collagen, acute UV irradiation transiently inhibits new synthesis of type I and type III procollagens [30]. Following UV irradiation of human skin in vivo, type I procollagen protein is significantly decreased within eight hours, is maximally reduced by approximately 70% by 24 hours, and returns to baseline levels within two to three days [30]. Loss of procollagen production is localized to the upper one-third of the dermis, reflecting the penetration of UV irradiation into the skin. At least two mechanisms contribute to downregulation of procollagen synthesis. First, as mentioned above, UV irradiation of human skin activates the transcription factor AP-1, which negatively regulates transcription of both genes that encode type I collagen chains, COL1A1 and COL1A2 [41, 42]. Second, UV irradiation impairs the signaling of

13

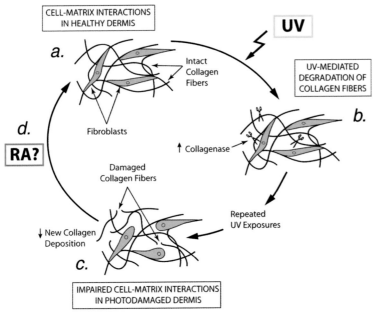

CELL-MATRIX INTERACTIONS
IN HEALTHY DERMIS

a.

Intact
Collagen
Fibers

UV

UV-MEDIATED
DEGRADATION OF
COLLAGEN FIBERS

b.

d.

RA?

Fibroblasts

↑ Collagenase

Damaged
Collagen Fibers

↓ New Collagen
Deposition

Repeated
UV Exposures

c.

IMPAIRED CELL-MATRIX INTERACTIONS
IN PHOTODAMAGED DERMIS

Fig.13.1. Mechanism for UV-mediated collagen depletion in human skin. In vivo and in healthy dermis, fibroblasts interact with intact collagen fibers at numerous adhesion points (*a*). UV irradiation increases collagenase, which degrades collagen fibers (*b*). With repeated UV exposures, damaged collagen fibers accumulate, thus impairing cell-matrix interactions (*c*). Impaired cell-matrix interactions are responsible for sustained decrease in collagen production in photodamaged skin. RA-mediated restoration of collagen formation (*d*) is the main underlying process in the repair of photodamaged skin

TGF-β, the major profibrotic cytokine, downregulating the type II TGF-β receptor and, to a lesser extent, increasing expression of the inhibitory regulator of the TGF-β signaling pathway, Smad7 [43, 44].

13.2.4 Cumulative Collagen Damage Contributes to the Phenotype of Photodamaged Skin

Photodamaged human skin is characterized by sustained downregulation of type I procollagen production, compared with matched individual sun-protected skin [45, 46]. Procollagen downregulation correlates with the clinical severity of photodamage [46], and therefore appears to reflect a substantial degree of accumulated damage.

Interestingly, the inherent capacity of fibroblasts in severely photodamaged human skin

to produce procollagen is not impaired. When grown in vitro, fibroblasts cultured from severely photodamaged skin and those cultured from sun-protected skin synthesize equal amounts of procollagen [47]. As described above, acute UV irradiation of skin induces transient activation of MMPs, with subsequent degradation of mature collagen. With repeated sun exposures over many years, fragmented collagen accumulates in photodamaged skin [47, 48]. In vitro treatment of collagen gels with MMP-1 generates collagen fragments that are similar to those observed in photodamaged human skin in vivo [48]. When dermal fibroblasts are seeded in a collagen gel pretreated with MMP-1, procollagen production is dramatically inhibited, compared to that from fibroblasts cultured on intact collagen gels [49]. Interestingly, removal of partially degraded collagen by treatment of the gels with MMP-9 restores procollagen biosynthesis. This loss of procollagen production results, at least in part, from reduced mechanical tension between the

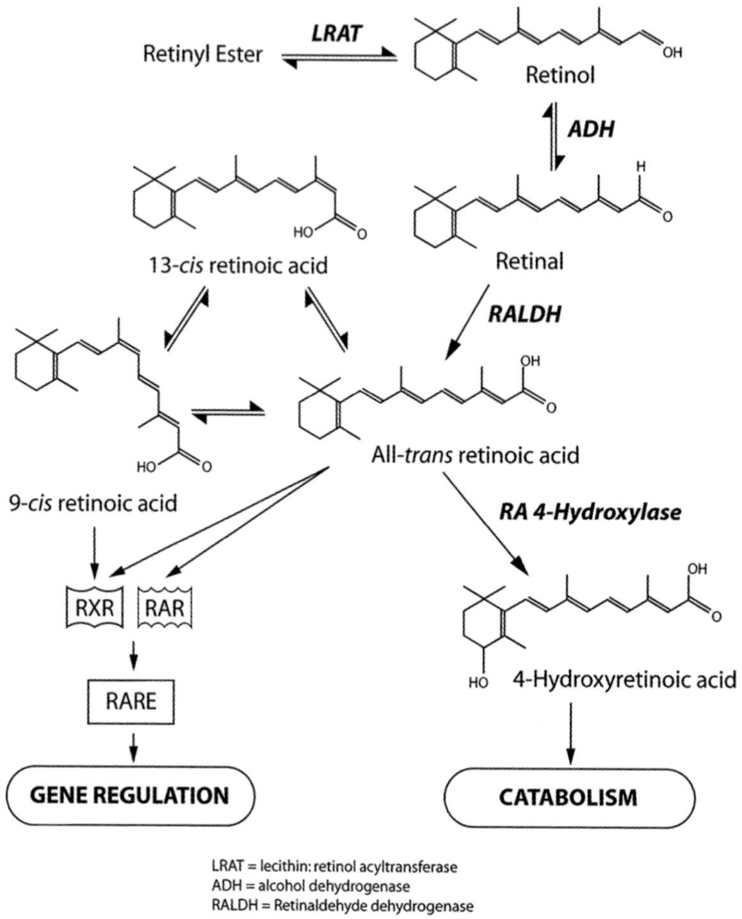

Fig. 13.2. Retinoid metabolism in the target cell

LRAT = lecithin: retinol acyltransferase
ADH = alcohol dehydrogenase
RALDH = Retinaldehyde dehydrogenase

13

fibroblasts and the partially degraded collagen gel.

Fibroblasts bind to collagen through specific receptors, integrins, which they display on their cell surface. Integrin binding allows fibroblasts to exert contractile forces on the collagen matrix, and in turn experience mechanical resistive forces exerted by the collagen matrix. Such mechanical tension has been demonstrated in model systems to influence a wide range of cellular behavior, including procollagen production by skin fibroblasts [50]. Therefore, it has been proposed [51] that accumulation of fragmented collagen in photodamaged skin weakens the resistive forces of the ECM, thereby reducing mechanical tension on fibroblasts, which in response downregulate procollagen synthesis (Fig. 13.1). This scenario describes a continual,

self-sustaining negative feedback pathway for loss of skin collagen, since UV exposure generates fragmented collagen by MMP-mediated degradation, and fragmented collagen inhibits procollagen production. In order to break this downward spiral, one must introduce a therapeutic agent that stimulates procollagen production in photodamaged skin. All-*trans*-RA has been demonstrated to be such an agent.

13.3 Retinoic Acid Metabolism in Skin

Retinoids constitute a group of natural and synthetic compounds characterized by vitamin A-like biological activity. Vitamin A (all-*trans*-retinol, ROL) and other natural retinoids, including all-*trans*-RA, play a critical role in embryogenesis, reproduction, vision, regulation of immunity, and epithelial cell differentiation. Retinoids are obtained from the diet, absorbed by the intestine, stored in the liver as retinyl esters (REs), and transported in the circulation to target cells as ROL. The mobilization of ROL from liver cells is regulated by the plasma concentration of unbound retinol binding protein (RBP). When ROL is released into the circulation, it binds RBP for delivery to target tissues. Within the target cell, ROL can be stored at the plasma membrane as REs, or converted into RA to exert biological effects. Conversion of ROL to RA consists of a two-step reaction (Fig. 13.2). First, ROL is converted into retinaldehyde (retinal, RAL). This rate-limiting step in RA metabolism is catalyzed by the enzyme alcohol dehydrogenase (ADH) [52]. The second step is the conversion of RAL into RA. This reaction, catalyzed by retinaldehyde dehydrogenase (RALDH), is irreversible. This is the reason why administration of RA to vitamin A-deficient animals does not increase RAL or ROL levels. In the cell, RA is not stored and its concentration is tightly controlled.

RA is bound within the cell cytoplasm by cellular RA binding protein (CRABP). RA bound to CRABP can undergo one of two fates: delivery to nuclear RA receptors (RARs), or catabolism into inactive forms. Catabolism is catalyzed by RA-4-hydroxylase, which belongs to the cytochrome P450 family [53]. RA concentrations in human skin are tightly regulated: topical application of RA to human skin in vivo increases RA degradation via induction of RA-hydroxylase [54], and decreases RA biosynthesis by increasing ROL-esterifying activity [55].

RA exerts its biological effects through binding to RARs, which belong to the superfamily of nuclear steroid hormone receptors. RARs partner with retinoid X receptors (RXR) to form functional ligand-activated transcription factors. RAR and RXR are each encoded by three separate genes, RAR-α, -β and -γ and RXR-α, -β and -γ. Among the different isoforms, adult human skin expresses predominantly RAR-γ and RXR-α [56, 57]. RA exists in three different stereoisomeric conformations: all-*trans*, 9-*cis*, and 13-*cis*. RARs bind all-*trans*-RA and 9-*cis*-RA; RXRs exclusively bind 9-*cis*-RA; neither RARs nor RXRs bind 13-*cis*-RA [58]. Upon binding of RA to RAR, RAR/RXR heterodimers form functional multimeric protein complexes that regulate transcription of target genes. Target genes contain specific DNA sequences in their promoter regions called RA response elements (RAREs) to which the RAR/RXR complexes bind [59].

13.4 Topical Retinoids Prevent Photoaging

Studies have shown that topical all-*trans*-RA can prevent UV-induced skin responses that lead to degradation and down-regulation of type I collagen in human skin in vivo. Pretreatment of human skin with RA inhibits UV-mediated induction of c-Jun protein and activation of AP-1 [28]. Inhibition of AP-1 by RA seems to occur through a mechanism distinct from transrepression [60, 61]. RA inhibits UV induction of c-Jun protein, but not c-Jun mRNA [28]. Whether RA-mediated inhibition of c-Jun protein induction results from decreased c-Jun protein synthesis and/or increased c-Jun protein degradation is not known.

RA inhibition of AP-1 activity results in reduction of UV-mediated induction of MMP-1, MMP-3, and MMP-9 mRNA and proteins [29], thereby preventing generation of fragmented collagen after UV exposure.

13.5 Topical Retinoids Repair Photoaging

13.5.1 Use of All-*Trans*-Retinoic Acid in the Treatment of Photoaging

In a pilot study, Kligman *et al.* found that all-*trans*-RA cream (tretinoin, Retin-A), which is used for the treatment of acne vulgaris, could partially reverse structural skin damage associated with photoaging [1]. In this vehicle-controlled, albeit unblinded study, the authors showed that 6 to 12 months of daily treatment with 0.05% tretinoin applied to the face and forearms induced a thickening of the atrophic epidermis, elimination of dysplasia and atypia, promoted a more uniform dispersion of melanin and neoformation of collagen and blood vessels. These observations were confirmed by several vehicle-controlled double-blind studies, the first of which was reported in 1988 by Weiss *et al.* This study demonstrated that 0.1% tretinoin cream used once daily for 4 months causes significant clinical improvement primarily in fine facial wrinkles, sallowness, looseness of the skin, and hyperpigmented macules (also called "liver spots" or actinic lentigines) [2]. Numerous subsequent clinical studies have confirmed these initial observations using 0.05% tretinoin for 3 to 6 months [3–6]. Interestingly, 0.1% and 0.025% tretinoin produce similar clinical and histological improvement of photoaging [7]. At a concentration of 0.02%, tretinoin is still effective for the treatment of photoaging [62], but lower doses (i.e. 0.01% and 0.005%) have no significant advantages over placebo [6]. When treatment was extended to a total of 9 months or more, persistence of clinical improvement in fine wrinkles and skin roughness has been reported [7, 63], even despite a reduction of dose or frequency of application after the first 4 months [64].

Higher doses of tretinoin are also effective in treating photoaging. In a double-blind placebo-controlled study, Rafal *et al.* found that 83% of patients treated with 0.1% tretinoin daily for 10 months, in contrast to 23% of placebo-treated patients, showed lightening of actinic lentigines. These results were confirmed by histology which showed a significant decrease in epidermal pigmentation in biopsy specimens. The improvement was sustained in subjects examined 6 months after the end of treatment [65]. This same treatment regimen also increased collagen skin content compared to vehicle-treated individuals [45].

In one study, long-term treatment was accompanied by loss of epidermal hyperplasia, suggesting that epidermal thickening is not directly related to clinical improvement [63]. In contrast, retinoid-mediated improvement of photoaging is associated with increased collagen I formation [45], reorganization of packed collagen fibers [66], and increased numbers of

type VII anchoring fibrils [67]. Solar elastosis is not substantially reduced; however, it tends to reside lower in the dermis, presumably due to accumulation of new collagen in the upper dermis, below the dermoepidermal junction.

Two observations regarding retinoid therapy of photoaged skin are very consistent. First, the magnitude of effects is largest in skin that originally shows the greatest degree of photodamage. Second, up to 92% of subjects using tretinoin in various studies report erythema and scaling at the site of application. This major side effect is referred to as "retinoid dermatitis", and can be a limitation to the use of retinoids for treatment of photoaging (see below).

13.5.2 Use of Other Retinoids in the Treatment of Photoaging

Because of the significant side effects associated with tretinoin therapy, the effectiveness of several other retinoids in the treatment of photoaged skin has been studied, ideally without causing irritation. For instance, 0.1% isotretinoin (13-*cis*-RA) cream used once daily for 8 months results in a reduction in wrinkles and actinic lentigines, accompanied by an increase in epidermal thickness [68]. However, significant erythema has been reported in 65% of subjects treated on the face with isotretinoin in contrast to 25% of subjects treated with vehicle. A statistically significant improvement of overall appearance of photodamaged skin has also been reported after 3 months of treatment with 0.05% isotretinoin followed by 0.1% for the next 6 months [69, 70].

Topical all-*trans*-RAL has been shown to increase epidermal thickness, increase differentiation markers and upregulate CRABP-II when used at 0.5%, 0.1% and 0.05% for 1 to 3 months in vivo [71]. RAL significantly reduces wrinkles, albeit to a lesser extent than tretinoin [72]. RAL also appears to be less irritating (23% of treated subjects) than tretinoin (32%).

ROL mimics the activity of 0.025% RA with respect to epidermal thickening when applied topically at 1.6% to human skin in vivo [73]. Like RA, ROL enhances expression of CRABP-

13

II, but is much less irritating than RA [73]. ROL and RAL are converted to RA by human keratinocytes, and this conversion is required for biological activity [74].

13.5.3 Retinoid Mechanism of Action

Despite extensive evidence demonstrating beneficial effects of topical retinoids in preventing and treating the clinical aspects of photodamaged skin, the detailed molecular basis of this activity remains elusive. Initially, irritation and scaling were thought to be the mechanism underlying retinoid-induced repair. However, several lines of evidence have dispelled this notion. First, two concentrations of tretinoin, 0.025% and 0.1%, were compared for their efficacy and irritancy [7]. Used once daily for 11 months, the two concentrations of tretinoin improved photoaged skin to similar extents, but 0.1% tretinoin was significantly more irritating. Second, topical retinyl palmitate (0.15%) induces skin irritation without demonstrating any advantages over placebo in treating photoaging [75]. In addition, non-retinoid agents that induce irritation and scaling do not enhance collagen synthesis in mouse skin [76]. These results demonstrate that irritation can be separated from efficacy, although irritation is an inherent side effect of retinoid therapy (see below).

The ability of retinoids to restore collagen formation is thought to be the main underlying mechanism by which the appearance of photodamaged skin is improved [45]. RA increases TGF-β in mouse skin [77, 78]. Since TGF-β is a major fibrotic cytokine, its induction by RA may underlie the ability of RA to induce collagen synthesis. However, the role of TGF-β in repair of photodamaged human skin remains to be investigated.

Studies of retinoid-mediated collagen induction in photodamaged human skin have been hampered by the lack of a suitable in vitro model. In vitro cultures of dermal fibroblasts constitutively produce high levels of collagen, which cannot be substantially increased by RA. As mentioned above, culturing fibroblasts in a collagen gel pretreated with human MMP-1 results in downregulation of collagen, as is observed in photodamaged skin [49]. Whether RA can restore collagen formation in this model remains to be determined.

Dermal ECM is a complex matrix composed of many structural components besides type I and type III collagens. Many ECM components have been reported to be altered in photodamaged skin, including elastin [79], anchoring fibrils, proteoglycans, and glycosaminoglycans [80]. It is possible that the mechanisms by which retinoids improve dermal ECM involves action on several dermal components in addition to collagen. For instance, RA increases fibrillin I, a component of microfibrils that are associated with elastic fibers [81]. RA also increases hyaluronic acid content in porcine skin [82] and in human skin organ culture [83]. Finally, it should be noted that, by stimulating collagen formation, RA would be expected to improve interactions between fibroblasts and ECM, which are impaired in photoaged dermis (Fig. 13.1).

13.6 Side Effects of Retinoids

Up to 92% of subjects who used tretinoin in various clinical studies have reported "retinoid dermatitis", i.e. erythema and scaling at the site of application [2]. The condition usually peaks early in therapy and disappears when treatment is discontinued. Scaling is the major deterrent to topical retinoid therapy and is often a limitation to the use of retinoids. Recent progress has been made in understanding the mechanisms by which retinoids induce epidermal hyperplasia.

Irritation may be explained by an overload of RA concentration in the epidermis. Epidermis is a non-homogeneous tissue comprising undifferentiated keratinocytes (basal layer) and differentiated cells (suprabasal layers). In normal human skin, the capacity of undifferentiated basal keratinocytes to esterify ROL is four times greater than that of differentiated suprabasal keratinocytes [55], and RA binding capacities increase with keratinocyte differentiation [84]. This scheme is consistent with the observation that a gradient of free RA influences keratinocyte differentiation and proliferation, with the concentration decreasing from the basal to the superficial layers [85]. When human skin

13

is treated with topical RA, the skin concentration of RA increases enough to activate gene transcription over retinoid receptors [86]. Exogenous RA enhances proliferation of basal keratinocytes [87], which induces an accelerated turnover of epidermal cells, and thickening of the epidermis [2, 88]. Retinoid treatment of normal human skin in vivo also decreases the cohesiveness of the stratum corneum, impairing the skin barrier and leading to an excessive scale.

RA-induced skin hyperplasia requires functional retinoid receptor [89, 90], and is primarily mediated by RARs [90–92]. The intensity of irritation and scaling is directly related to the amount of RA administered to the skin, since lowering the concentration of topical RA reduces irritation, but does not impair biological activity [7].

The involvement of EGF-R (ErbB1) in retinoid-mediated epidermal hyperplasia has been demonstrated in several in vivo and in vitro studies. RA-induced epidermal hyperplasia is triggered by increased secretion of two EGF-R ligands, heparin-binding epidermal growth factor (HB-EGF) and amphiregulin, in suprabasal cells of normal epidermis in vivo (Rittié et al., in preparation). Blockage of HB-EGF or amphiregulin with antibodies (Rittié et al., in preparation) or inhibition of EGF-R phosphorylation by specific inhibitor [87] strongly reduces the RA-induced epidermal hyperplasia in human skin organ cultures. In vivo, topical application of genistein, an inhibitor of tyrosine kinase compatible with topical use, blocks retinoid-induced hyperproliferation of epidermis in human skin (Rittié et al., in preparation). Interestingly, inhibition of retinoid-associated hyperplasia does not impair the ability of retinoid to increase collagen synthesis by fibroblasts in skin organ culture [87]. It would be interesting to determine whether the separation of these two features can be achieved in vivo for the treatment of photoaging.

Topical application of RA precursors causes less irritation than application of RA itself. These precursors do not bind retinoid receptors and must be converted within the cell into RA to exert biological activity [74]. Interestingly, the extent of irritation is inversely proportional to the number of steps required for conversion of

the precursor into RA. For instance, 13-*cis*-RA is less irritating than all-*trans*-RA [68], probably because it needs to be isomerized into 9-*cis*-RA or all-*trans*-RA in order to bind retinoid receptors [58]. Moreover, RA has been shown to be more irritating than RAL [72], which is more irritating than ROL [93], the conversion of RAL and ROL into RA being a one- and two-step process, respectively (Fig. 13.2). Failure to detect significant accumulation of RA in ROL-treated skin [73] shows that RA synthesis is tightly regulated, and RA is rapidly used and degraded. Overloading of RA in skin cells is likely to induce skin irritation.

Concern has been voiced that repeated use of topical tretinoin might increase RA plasma concentrations. This issue was carefully addressed and rejected in different studies showing that topical tretinoin treatment has no effect on endogenous plasma levels of RA or its metabolites [68, 94]. Percutaneously absorbed tretinoin represents approximately 1–2% of a single topical dose applied daily for up to one year [94]. Therefore, 1 ml of 0.05% (500 ng/ml) tretinoin applied topically would provide a maximum of 10 ng RA in the bloodstream (retinoid plasma concentration of 1–4 µg/l). Another study has shown showed that daily application of 0.1% tretinoin for 8 months to the face, forearms and hands does not significantly increase plasma retinoid levels [68]. This is certainly the reason why topical tretinoin is not associated with a detectable risk of major congenital disorders [95], as opposed to oral retinoids that are teratogenic when administered during early pregnancy [96].

13.7 Conclusions

Photoaging is caused by the skin's reaction to chronic sun exposure. Topical retinoids are effective in treating clinical signs of photoaging, including fine wrinkles, coarseness and laxity. Retinoid dermatitis is inherent to retinoid therapy, since it is mediated by nuclear retinoid receptors. Irritation can be minimized by reducing dose and frequency of treatment. Understanding the molecular mechanisms of action of retinoids in photoaging will provide pharmacological targets for the development of non-retinoid compounds that have the beneficial properties of retinoids, without the unwanted side effects.

Acknowledgements

During the writing of this chapter, Laure Rittié was supported by a grant from Pfizer, Inc.

References

1. Kligman A, Grove G, Hirose R, Leyden J (1986) Topical tretinoin for photoaged skin. J Am Acad Dermatol 15:836–859
2. Weiss J, Ellis C, Headington J, Tincoff T, Hamilton T, Voorhees J (1988) Topical tretinoin improves photoaged skin. JAMA 259:527–532
3. Leyden J, Grove G, Grove M, Thorne E, Lufrano L (1989) Treatment of photodamaged facial skin with topical tretinoin. J Am Acad Dermatol 21:638–644
4. Lever L, Kumar P, Marks R (1990) Topical retinoic acid for treatment of solar damage. Br J Dermatol 122:91–98
5. Weinstein G, Nigra T, Pochi P, Savin R, Allan A, Benik K, et al (1991) Topical tretinoin for treatment of photodamaged skin. A multicenter study. Arch Dermatol 127:659–665
6. Olsen E, Katz H, Levine N, Shupack J, Billys M, Prawer S, et al (1992) Tretinoin emollient cream: a new therapy for photodamaged skin. J Am Acad Dermatol 26:215–224
7. Griffiths C, Kang S, Ellis C, Kim K, Finkel L, Ortiz-Ferrer L, et al (1995) Two concentrations of topical tretinoin (retinoic acid) cause similar improvement of photoaging but different degrees of irritation: a double-blind, vehicle-controlled comparison of 0.1% and 0.025% tretinoin creams. Arch Dermatol 131:1037–1044
8. Lavker R (1995) Cutaneous aging: chronologic versus photoaging. In: Gilchrest B (ed) Photodamage, vol 1. Blackwell, Cambridge, pp 123–135
9. Hanson K, Clegg R (2002) Observation and quantification of ultraviolet-induced reactive oxygen species in ex vivo human skin. Photochem Photobiol 76:57–63
10. McCormick J, Fischer J, Pachlatko J (1976) Characterization of a cell-lethal product from the photooxidation of tryptophan: hydrogen peroxide. Science 191:468–469
11. Dalle Carbonare M, Pathak M (1992) Skin photosensitizing agents and the role of reactive oxygen species in photoaging. J Photochem Photobiol B 14:105–124
12. Hanson K, Simon J (1998) Epidermal trans-urocanic acid and the UV-A-induced photoaging of the skin. Proc Natl Acad Sci U S A 95:10576–10578
13. Wondrak G, Roberts M, Cervantes-Laurean D, Jacobson M, Jacobson E (2003) Proteins of the extracellular matrix are sensitizers of photo-oxidative stress in human skin cells. J Invest Dermatol 121:578–586
14. Wondrak G, Roberts M, Jacobson M, Jacobson E (2002) Photosensitized growth inhibition of cultured human skin cells: mechanisms and suppression of oxidative stress from solar irradiation of glycated proteins. J Invest Dermatol 119:489–498
15. Masaki H, Okano Y, Sakurai H (1997) Generation of active oxygen species from advanced glycation end-products (AGE) under ultraviolet light A (UVA) irradiation. Biochem Biophys Res Commun 235:306–310
16. Yasui H, Sakurai H (2000) Chemiluminescent detection and imaging of reactive oxygen species in live mouse skin exposed to UVA. Biochem Biophys Res Commun 269:131–136
17. Warmuth I, Harth Y, Matsui M, Wang N, DeLeo V (1994) Ultraviolet radiation induces phosphorylation of the epidermal growth factor receptor. Cancer Res 54:374–376
18. Sachsenmaier C, Radler-Pohl A, Zinck R, Nordheim A, Herrlich P, Rahmsdorf H (1994) Involvement of growth factor receptors in the mammalian UVC response. Cell 78:963–972
19. Coffer P, Burgering B, Peppelenbosch M, Bos J, Kruijer W (1995) UV activation of receptor tyrosine kinase activity. Oncogene 11:561–569
20. Knebel A, Rahmsdorf H, Ullrich A, Herrlich P (1996) Dephosphorylation of receptor tyrosine kinases as target of regulation by radiation, oxidants or alkylating agents. EMBO J 15:5314–5325
21. Rosette C, Karin M (1996) Ultraviolet light and osmotic stress: activation of the JNK cascade through multiple growth factor and cytokine receptors. Science 274:1194–1197
22. Fisher G, Voorhees J (1998) Molecular mechanisms of photoaging and its prevention by retinoic acid: ultraviolet irradiation induces MAP kinase signal transduction cascades that induce AP-1-regulated matrix metalloproteinases that degrade human skin in vivo. J Invest Dermatol Symp Proc 3:61–68
23. Persson C, Sjöblom T, Groen A, Kappert K, Engström U, Hellman U, et al (2004) Preferential oxidation of the second phosphatase domain of receptor-like PTP-α revealed by an antibody against oxidized protein tyrosine phosphatases. Proc Natl Acad Sci U S A 101:1886–1891
24. Meng T-C, Fukada T, Tonks N (2002) Reversible oxidation and inactivation of protein tyrosine phosphatases in vivo. Mol Cell 9:387–399

25. Denu J, Tanner K (1998) Specific and reversible inactivation of protein tyrosine phosphatases by hydrogen peroxide: evidence for a sulfenic acid intermediate and implications for redox regulation. Biochemistry 37:5633–5642

26. Gulati P, Markova B, Göttlicher M, Böhmer F, Herrlich P (2004) UVA inactivates protein tyrosine phosphatases by calpain-mediated degradation. EMBO Rep 5:812–817

27. Fisher G, Kang S, Varani J, Bata-Csorgo Z, Wan Y, Datta S, et al (2002) Mechanisms of photoaging and chronological skin aging. Arch Dermatol 138:1462–1470

28. Fisher G, Talwar H, Lin J, McPhillips F, Wang Z, Li X, et al (1998) Retinoic acid inhibits induction of c-Jun protein by ultraviolet radiation that occurs subsequent to activation of mitogen-activated protein kinase pathways in human skin in vivo. J Clin Invest 101:1432–1440

29. Fisher G, Datta S, Talwar H, Wang Z, Varani J, Kang S, et al (1996) Molecular basis of sun-induced premature skin ageing and retinoid antagonism. Nature 379:335–339

30. Fisher G, Datta S, Wang Z, Li X, Quan T, Chung J, et al (2000) c-Jun-dependent inhibition of cutaneous procollagen transcription following ultraviolet irradiation is reversed by all-trans retinoic acid. J Clin Invest 106:663–670

31. Fisher G, Wang Z, Datta S, Varani J, Kang S, Voorhees J (1997) Pathophysiology of premature skin aging induced by ultraviolet light. New Engl J Med 337:1419–1428

32. Kang S, Chung J, Lee J, Fisher G, Wan Y, Duell E, et al (2003) Topical N-acetyl cysteine and genistein prevent ultraviolet-light-induced signaling that leads to photoaging in human skin in vivo. J Invest Dermatol 120:835–841

33. Bender K, Göttlicher M, Whiteside S, Rahmsdorf H, Herrlich P (1998) Sequential DNA damage-independent and -dependent activation of NF-κB by UV. EMBO J 17:5170–5181

34. Li N, Karin M (1998) Ionizing radiation and short wavelength UV activate NF-κB through two distinct mechanisms. Proc Natl Acad Sci U S A 95:13012–13017

35. Wu S, Tan M, Hu Y, Wang J-L, Scheuner D, Kaufman R (2004) Ultraviolet light activates NFκB through translational inhibition of IκBα synthesis. J Biol Chem 279:34898–34902

36. Huang T, Feinberg S, Suryanarayanan S, Miyamoto S (2002) The zinc finger domain of NEMO is selectively required for NF-κB activation by UV radiation and topoisomerase inhibitors. Mol Cell Biol 22:5813–5825

37. Park K-J, Gaynor R, Kwak Y (2003) Heat shock protein 27 association with the IκB kinase complex regulates tumor necrosis factor α-induced NF-κB activation. J Biol Chem 278:35272–35278

38. Yamamoto Y, Gaynor R (2001) Therapeutic potential of inhibition of the NF-κB pathway in the treatment of inflammation and cancer. J Clin Invest 107:135–142

39. Csiszar K (2001) Lysyl oxidases: a novel multifunctional amine oxidase family. Prog Nucleic Acid Res Mol Biol 70:1–32

40. Sternlicht M, Werb Z (2001) How matrix metalloproteinases regulate cell behavior. Annu Rev Cell Dev Biol 17:463–516

41. Liska D, Slack J, Bornstein P (1990) A highly conserved intronic sequence is involved in transcriptional regulation of the α1(I) collagen gene. Cell Regul 1:487–498

42. Chung K-Y, Agarwal A, Uitto J, Mauviel A (1996) An AP-1 binding sequence is essential for regulation of the human α2(I) collagen promoter activity by transforming growth factor-β. J Biol Chem 271:3272–3278

43. Quan T, He T, Voorhees J, Fisher G (2001) Ultraviolet irradiation blocks cellular responses to transforming growth factor-β by down-regulating its type-II receptor and inducing Smad7. J Biol Chem 276:26349–26356

44. Quan T, He T, Kang S, Voorhees J, Fisher G (2004) Solar ultraviolet irradiation reduces collagen in photoaged human skin by blocking transforming growth factor-β type II receptor/Smad signaling. Am J Pathol 165:741–751

45. Griffiths C, Russman A, Majmudar G, Singer R, Hamilton T, Voorhees J (1993) Restoration of collagen formation in photodamaged human skin by tretinoin (retinoic acid). New Engl J Med 329:530–535

46. Talwar H, Griffiths C, Fisher G, Hamilton T, Voorhees J (1995) Reduced type I and type III procollagens in photodamaged adult human skin. J Invest Dermatol 105:285–290

47. Varani J, Spearman D, Perone P, Fligiel S, Datta S, Wang Z, et al (2001) Inhibition of type I procollagen synthesis by damaged collagen in photoaged skin and by collagenase-degraded collagen in vitro. Am J Pathol 158:931–942

48. Fligiel S, Varani J, Datta S, Kang S, Fisher G, Voorhees J (2003) Collagen degradation in aged/photodamaged skin in vivo and after exposure to matrix metalloproteinase-1 in vitro. J Invest Dermatol 120:842–848

49. Varani J, Perone P, Fligiel S, Fisher G, Voorhees J (2002) Inhibition of type I procollagen production in photodamage: correlation between presence of high molecular weight collagen fragments and reduced procollagen synthesis. J Invest Dermatol 119:122–129

13

50. Mauch C, Hatamochi A, Scharffetter K, Krieg T (1988) Regulation of collagen synthesis in fibroblasts within a three-dimensional collagen gel. Exp Cell Res 178:493–503

51. Varani J, Scuger L, Dame M, Leonard C, Fligiel S, Kang S, et al (2004) Reduced fibroblast interaction with intact collagen as a mechanism for depressed collagen synthesis in photodamaged skin. J Invest Dermatol 122:1471–1479

52. Roos T, Jugert F, Merk H, Bickers D (1998) Retinoid metabolism in the skin. Pharmacol Rev 50:315–333

53. White J, Guo Y-D, Baetz K, Beckett-Jones B, Bonasoro J, Hsu K, et al (1996) Identification of the retinoic acid-inducible all-trans-retinoic acid 4-hydroxylase. J Biol Chem 271:29922–29927

54. Duell E, Kang S, Voorhees J (1996) Retinoic acid isomers applied to human skin in vivo each induce a 4-hydroxylase that inactivates only trans retinoic acid. J Invest Dermatol 106:316–320

55. Kurlandsky S, Duell E, Kang S, Voorhees J, Fisher G (1996) Auto-regulation of retinoic acid biosynthesis through regulation of retinol esterification in human keratinocytes. J Biol Chem 271:15346–15352

56. Fisher G, Talwar H, Xiao J-H, Datta S, Reddy A, Gaub M-P, et al (1994) Immunological identification and functional quantification of retinoic acid and retinoid X receptor proteins in human skin. J Biol Chem 269:20629–20635

57. Elder J, Fisher G, Zhang Q-Y, Eisen D, Krust A, Kastner P, et al (1991) Retinoic acid receptor gene expression in human skin. J Invest Dermatol 96:425–433

58. Allenby G, Bocquel M-T, Saunders M, Kazmer S, Speck J, Rosenberger M, et al (1993) Retinoic acid receptors and retinoid X receptors: interactions with endogenous retinoic acids. Proc Natl Acad Sci U S A 90:30–34

59. Chambon P (1994) The retinoid signaling pathway: molecular and genetic analyses. Semin Cell Biol 5:115–125

60. Chen S, Ostrowski J, Whiting G, Roalsvig T, Hammer L, Currier S, et al (1995) Retinoic acid receptor gamma mediates topical retinoid efficacy and irritation in animal models. J Invest Dermatol 104:779–783

61. DiSepio D, Sutter M, Johnson A, Chandraratna R, Nagpal S (1999) Identification of the AP1-antagonism domain of retinoic acid receptors. Mol Cell Biol Res Commun 1:7–13

62. Nyirady J, Bergfeld W, Ellis C, Levine N, Savin R, Shavin J, et al (2001) Tretinoin cream 0.02% for the treatment of photodamaged facial skin: a review of 2 double-blind clinical studies. Cutis 68:135–142

63. Bhawan J, Olsen E, Lufrano L, Thorne E, Schwab B, Gilchrest B (1996) Histologic evaluation of the long term effects of tretinoin on photodamaged skin. J Dermatol Sci 11:177–182

64. Ellis C, Weiss J, Hamilton T, Headington J, Zelickson A, Voorhees J (1990) Sustained improvement with prolonged topical tretinoin (retinoic acid) for photoaged skin. J Am Acad Dermatol 23:629–637

65. Rafal E, Griffiths C, Ditre C, Finkel L, Hamilton T, Ellis C, et al (1992) Topical tretinoin (retinoic acid) treatment for liver spots associated with photodamage. N Engl J Med 326:368–374

66. Yamamoto O, Bhawan J, Solares G, Tsay A, Gilchrest B (1995) Ultrastructural effects of topical tretinoin on dermo-epidermal junction and papillary dermis in photodamaged skin. A controlled study. Exp Dermatol 4:146–154

67. Chen M, Goyal S, Cai X, O'Toole E, Woodley D (1997) Modulation of type VII collagen (anchoring fibril) expression by retinoids in human skin cells. Biochim Biophys Acta 1351:333–340

68. Maddin S, Lauharanta J, Agache P, Burrows L, Zultak M, Bulger L (2000) Isotretinoin improves the appearance of photodamaged skin: results of a 36-week, multicenter, double-blind, placebo-controlled trial. J Am Acad Dermatol 42:56–63

69. Armstrong R, Lesiewicz J, Harvey G, Lee L, Spoehr K, Zultak M (1992) Clinical panel assessment of photodamaged skin treated with isotretinoin using photographs. Arch Dermatol 128:352–356

70. Sendagorta E, Lesiewicz J, Armstrong R (1992) Topical isotretinoin for photodamaged skin. J Am Acad Dermatol 27:S15–S18

71. Saurat J, Didierjean L, Masgrau E, Piletta P, Jaconi S, Chatellard-Gruaz D, et al (1994) Topical retinaldehyde on human skin: biologic effects and tolerance. J Invest Dermatol 103:770–774

72. Creidi P, Vienne M-P, Ochonisky S, Lauze C, Turlier V, Lagarde J-M, et al (1998) Profilometric evaluation of photodamage after topical retinaldehyde and retinoic acid treatment. J Am Acad Dermatol 39:960–965

73. Kang S, Duell E, Fisher G, Datta S, Wang Z, Reddy A, et al (1995) Application of retinol to human skin in vivo induces epidermal hyperplasia and cellular retinoid binding proteins characteristic of retinoic acid but without measurable retinoic acid levels or irritation. J Invest Dermatol 105:549–556

74. Kurlandsky S, Xiao J-H, Duell E, Voorhees J, Fisher G (1994) Biological activity of all-trans retinol requires metabolic conversion to all-trans retinoic acid and is mediated through activation of nuclear retinoid receptors in human keratinocytes. J Biol Chem 269:32821–32827

75. Green C, Orchard G, Cerio R, Hawk J (1998) A clinicopathological study of the effects of topical retinyl propionate cream in skin photoageing. Clin Exp Dermatol 23:162–167

76. Kligman L, Sapadin A, Schwartz E (1996) Peeling agents and irritants, unlike tretinoin, do not stimu-

13

late collagen synthesis in the photoaged hairless mouse. Arch Dermatol Res 288:615–620

77. Kim H, Bogdan N, D'Agostaro L, Gold L, Bryce G (1992) Effect of topical retinoic acids on the levels of collagen mRNA during the repair of UVB-induced dermal damage in the hairless mouse and the possible role of TGF-β as a mediator. J Invest Dermatol 98:359–363

78. Glick A, Flanders K, Danielpour D, Yuspa S, Sporn M (1989) Retinoic acid induces transforming growth factor-β2 in cultured keratinocytes and mouse epidermis. Cell Regul 1:87–97

79. Mitchell R (1967) Chronic solar dermatosis: a light and electron microscopic study of the dermis. J Invest Dermatol 48:203–220

80. Bernstein E, Underhill C, Hahn P, Brown D, Uitto J (1996) Chronic sun exposure alters both the content and distribution of dermal glycosaminoglycans. Br J Dermatol 135:255–262

81. Watson R, Craven N, Kang S, Jones C, Kielty C, Griffiths C (2001) A short-term screening protocol, using fibrillin-1 as a reporter molecule, for photoaging repair agents. J Invest Dermatol 116:672–678

82. King I (1984) Increased epidermal hyaluronic acid synthesis caused by four retinoids. Br J Dermatol 110:607–608

83. Tammi R, Ripellino J, Margolis R, Maibach H, Tammi M (1989) Hyaluronate accumulation in human epidermis treated with retinoic acid in skin organ culture. J Invest Dermatol 92:326–332

84. Elder J, Åström A, Pettersson U, Tavakkol A, Griffiths C, Krust A, et al (1992) Differential regulation of retinoic acid receptors and binding proteins in human skin. J Invest Dermatol 98:673–679

85. Darmon M, Blumenberg M (1993) Retinoic acid in epithelial and epidermal differentiation. In: Darmon M, Blumenberg M (eds) Molecular biology of the skin. Academic Press, San Diego, pp 181–206

86. Duell E, Åström A, Griffiths C, Chambon P, Voorhees J (1992) Human skin levels of retinoic acid and cytochrome P-450-derived 4-hydroxyretinoic acid after topical application of retinoic acid in vivo compared to concentrations required to stimulate retinoic acid receptor-mediated transcription in vitro. J Clin Invest 90:1269–1274

87. Varani J, Ziegler M, Dame M, Kang S, Fisher G, Voorhees J, et al (2001) Heparin-binding epidermal-growth factor-like growth factor activation of keratinocyte ErbB receptors mediates epidermal hyperplasia, a prominent side-effect of retinoid therapy. J Invest Dermatol 117:1335–1341

88. Fisher G, Esmann J, Griffiths C, Talwar H, Duell E, Hammerberg C, et al (1991) Cellular, immunologic and biochemical characterization of topical retinoic acid-treated human skin. J Invest Dermatol 96:699–707

89. Chapellier B, Mark M, Messaddeq N, Calléja C, Warot X, Brocard J, et al (2002) Physiological and retinoid-induced proliferations of epidermis basal keratinocytes are differently controlled. EMBO J 21:3402–3413

90. Xiao J-H, Feng X, Di W, Peng Z-H, Li L-A, Chambon P, et al (1999) Identification of heparin-binding EGF-like growth factor as a target in intercellular regulation of epidermal basal cell growth by suprabasal retinoic acid receptors. EMBO J 18:1539–1548

91. Chen J, Penco S, Ostrowski J, Balaguer P, Pons M, Starrett J, et al (1995) RAR-specific agonist/antagonists which dissociate transactivation and AP1 transrepression inhibit anchorage-independent cell proliferation. EMBO J 14:1187–1197

92. Thacher S, Standeven A, Athanikar J, Kopper S, Castilleja O, Escobar M, et al (1997) Receptor specificity of retinoid-induced epidermal hyperplasia: effect of RXR-selective agonists and correlation with topical irritation. J Pharmacol Exp Ther 282:528–534

93. Fluhr J, Vienne M, Laure C, Dupuy P, Gehring W, Gloor M (1999) Tolerance profile of retinol, retinaldehyde and retinoic acid under maximized and long-term clinical conditions. Dermatology 199:57–60

94. Latriano L, Tzimas G, Wong F, Willis R (1997) The percutaneous absorption of topically applied tretinoin and its effect on endogenous concentrations of tretinoin and its metabolites after single doses or long-term use. J Am Acad Dermatol 36:S77–S85

95. Jick S, Terris B, Jick H (1993) First trimester topical tretinoin and congenital disorders. Lancet 341:1181–1182

96. Lipson A, Collins F, Webster W (1993) Multiple congenital defects associated with maternal use of topical tretinoin. Lancet 341:1352–1353

Chapter 14

Topical Photodynamic Therapy

Dany J. Touma, Rolf-Markus Szeimies

14

Contents

14.1 Introduction

At the beginning of the 20th century Hermann von Tappeiner, director of the Institute of Pharmacology at the University of Munich, coined the term "photodynamic reaction" to describe the oxygen-dependent tissue reaction following photosensitization and irradiation with light [1]. Today it is known that photodynamic therapy (PDT) requires the simultaneous presence of a photosensitizer, light and oxygen in the diseased tissue. The photosensitizer accumulates in the target cells and absorbs light of a certain wavelength. The energy is transferred to oxygen and highly reactive oxygen species (ROS) – mainly singlet oxygen – are generated. Following an appropriate light dose the reactive oxygen species directly lead to cell and tissue damage by inducing necrosis and apoptosis and indirectly stimulate inflammatory cell mediators (Fig. 14.1).

In recent decades, PDT has gained worldwide popularity, first as an experimental therapy for a variety of human cancers. Mainly porphyrins, chlorin derivatives or phthalocyanines have been studied so far for primary and adjuvant cancer therapy [2]. However, for dermatological purposes, only hematoporphyrin derivatives (HPD) such as porfimer sodium (Photofrin) and porphyrin-inducing precursors such as 5-aminolevulinic acid (ALA) and methyl aminolevulinate (MAL) are of practical use. As systemic photosensitizing drugs induce prolonged phototoxicity [3], topical photosensitizers are preferred for use in dermatology. Meanwhile drugs such as ALA and MAL have reached approval status for the treatment of epithelial cancers or their precursors throughout the world and there is growing interest in the use of PDT not only for nonmelanoma skin cancer but also

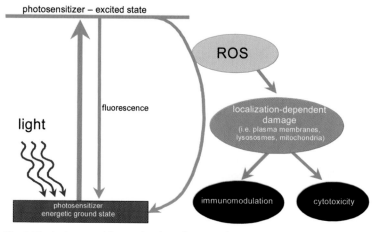

Fig. 14.1. A photosensitizer molecule in the excited state is able to form ROS, mainly singlet oxygen (reaction type II) via photooxidation. Depending on the subcellular location of the dye, organelle-specific damage occurs. The specific damage sites plus the extent of damage lead to immunomodulation and/or cytotoxicity (from Szeimies RM et al. (2001) PDT in Dermatology in: Krutmann J et al. Dermatological Phototherapy and Photodiagnostic Methods, Springer)

for other skin tumors such as lymphoma or for tumor surveillance in transplant patients as well as for non-oncological indications such as psoriasis, localized scleroderma and skin rejuvenation [4–6].

14.2 Mechanism of Action

During irradiation the photosensitizer absorbs light and this is converted to an energetically higher state, the "singlet state". After the short half-life period (approximately 10^{-9} s), the activated photosensitizer returns to the ground state after emission of fluorescence and/or internal conversion, or changes from the singlet state into the more stable triplet state with a longer half-life (10^{-3} s) ("intersystem crossing"). In the type I photooxidative reaction there is direct hydrogen and electron transfer from the photosensitizer in the triplet state to a substrate. This reaction results in the generation of radicals of the substrate. These radicals are able to react directly with molecular oxygen to form peroxides, hydroxy radicals and superoxide anions. This type I reaction is strongly concentration-dependent. Cells can be directly damaged by this reaction, especially when the photosensitizer is bound to easily oxidizable molecules.

In the photooxidative type II reaction, electrons or energy are directly transferred to molecular oxygen in the ground state (triplet) and singlet oxygen is formed. The highly reactive state of singlet oxygen results in a very effective oxidation of biological substrates. Both reaction types can occur in parallel, as substrate and molecular oxygen are present together with the photosensitizer in the triplet state. The type of reaction that preferentially occurs depends on the photosensitizer used, the subcellular location of the dye and the substrate, and the oxygen supply around the activated photosensitizer. Indirect experiments in vitro have indicated that singlet oxygen is the main mediator of PDT-induced biological effects.

Following activation of a photosensitizer with light of the appropriate wavelength ROS, in particular singlet oxygen, are generated. Depending on the amount and location in the target tissue, these ROS modify either cellular functions or induce cell death by necrosis or apoptosis [2, 7] (Fig. 14.2).

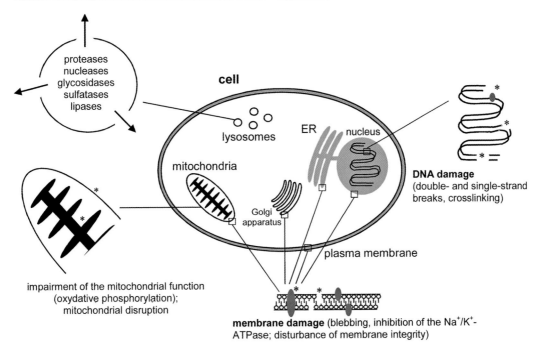

Fig. 14.2. PDT-related subcellular damages depend on the localization of the photosensitizer used. Significant damage occurs mostly at the plasma membranes, lysosomes and mitochondria. This results in disturbance of the membrane integrity, the release of lysosomal enzymes and impairment of the respiratory chain. Damage to the DNA does not contribute significantly to cellular necrosis (from Ref. 7)

14.3 Photosensitizers

Topically applied dyes including eosin red and erythrosine were the first "photosensitizers" used to treat conditions such as pityriasis versicolor, psoriasis, molluscum contagiosum, syphilis, lupus vulgaris and skin cancer [1]. The tumor localizing effects of porphyrins have been studied since 1908. In the late 1970s Dougherty used HPD to treat skin cancer [1, 2], and this resulted in a renewal of interest in PDT. The main problem in the use of HPD is the prolonged skin photosensitization which lasts for several weeks [8]. Topical application of these drugs is not possible since the rather large molecules (tetrapyrrol rings) do not penetrate the skin. Therefore the introduction of porphyrin precursors such as ALA and MAL by Kennedy and coworkers in 1990 was a significant milestone in the development of PDT in dermatology as the small low molecular weight molecules can easily penetrate the epidermis [2]. Currently in Europe

MAL is approved under the name of Metvix for the treatment of basal cell carcinoma and actinic keratoses (AK) in combination with red light. It is also registered in the US for AK under the name Metvixia but is not yet marketed. In the U.S., Metvixia® is approved for treating AKs since 2004, while 5-ALA hydrochloride (Levulan® Kerastick™) is approved for photodynamic treatment of AKs in combination with blue light since 1999 [3]. Photosensitisers based on 5-ALA are not photoactive by themselves, but show a preferential intracellular accumulation in the altered cells of the diseased tissue and are metabolized during haem biosynthesis to photosensitizing porphyrins quite selectively in these cells [7, 9]. If no surface illumination is given, the photoactive porphyrins are metabolized to the photodynamically inactive haem within the following 24 to 48 hours [2].

Since proliferating, relatively iron-deficient tumor cells of epithelial origin are highly sensitized by ALA and MAL, tissue damage is mostly

14

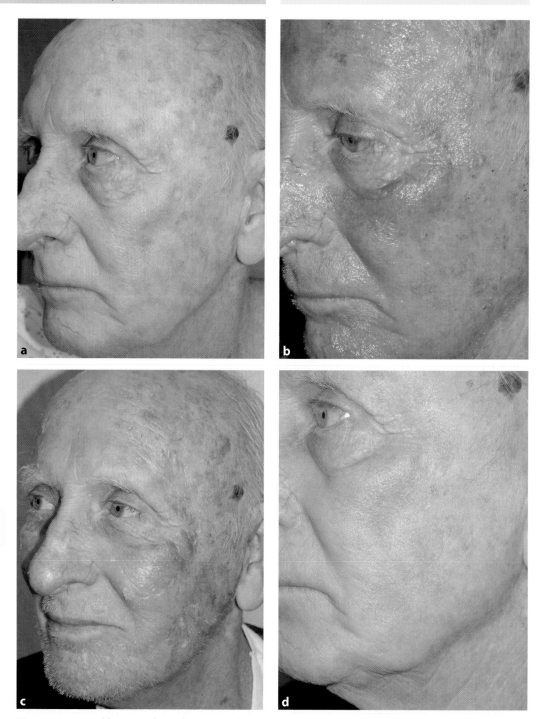

Fig. 14.3. 73-year-old patient who underwent ALA-PDT (Levulan Kerastick) for actinic keratoses: **a** pretreatment; **b** 48 hours after treatment; **c** 1 week after treatment; **d** 1 month after treatment. Note the significant improvement in skin texture and discoloration

restricted to the sensitized cells, almost omitting the surrounding tissue, especially cells of mesenchymal origin such as fibroblasts, resulting in excellent cosmesis (Fig. 14.3) [9]. Aside from two case reports which may be no more than coincidence, no reports on the carcinogenicity of ALA/MAL-PDT have been published [9]. Moreover, in a recent study even long-term topical application of ALA and subsequent irradiation with blue light in a hairless mouse model did not induce skin tumors [10]. Stender et al. have even shown a delay in photoinduced carcinogenesis in mice following repetitive treatments with ALA-PDT [11].

14.4 Light Sources Used in Topical PDT

Protoporphyrin IX (PPIX), the leading porphyrin induced after ALA or MAL sensitization is maximally activated at 409 nm, in the Soret band area of the spectrum, with significantly lower peaks at 509, 544, 584, and 635 nm (Q bands). The light sources used for PDT include red light sources (metal halide, LED, xenon lamps with cut-off filters), argon lasers, simple slide projectors or other broadband light devices, intense pulse light (IPL) lasers, and pulsed dye lasers. Blue light, with a peak wavelength of 417 nm (range 412–422 nm) corresponds to the area of maximal PPIX light absorption, and provides the most effective light activation of PPIX. A device delivering blue light (BLU-U, DUSA Pharmaceuticals, Wilmington, Mass.) has the additional advantage of a large field diameter that allows treatment of broad areas such as the full face and scalp. Blue light PDT has been shown to be effective for nonhypertrophic AK at the low fluence of 10 J/cm^2, requiring approximately 16 min 40 s of light exposure time. Using this light source with ALA, an incubation time of 14–18 hours is currently the only PDT method approved by the FDA for the treatment of AK. Blue light is believed to penetrate less than 2 mm into skin, and therefore should be restricted in use to the treatment of AK only [12].

Longer wavelengths of light, such as red light, are desirable when deeper parts or thicker lesions as in Bowen's disease or basal cell carcinoma are being targeted, but then higher light doses (usually 75–100 J/cm^2, depending on the bandwith of the light source) are needed to compensate for the porphyrin's lower absorption coefficient in that wavelength range. IPL, which provides a range of wavelengths of light, and flashlamp-pumped pulsed dye lasers are also suitable for PDT since they emit light in the activating range [13]. The advantages of these light sources over blue light are a greater time efficiency, the possibility of stacking pulses in order to intensify the treatment, and the improvement of associated vascular and pigmented lesions in broad facial treatments. However, they are more expensive and require technical skill.

In summary, blue light is the most potent light source for activation of ALA-induced porphyrins but restricted to superficial lesions. IPL and pulse dye lasers are expensive and technique-sensitive, but they are time-efficient and useful for stacking pulses to thicker individual lesions.

14.5 Topical PDT for Actinic Keratoses

The affinity of topical PDT for dysplastic skin has made it an immediate research tool in the area of AK. Indeed, topical PDT has been shown in several studies to be highly effective for the treatment of AKs. In 1999 the FDA approved topical PDT as safe and efficacious for the spot treatment of AKs as an alternative to liquid nitrogen cryotherapy [14, 15]. The FDA-approved protocol specifies a 14–18-hour delay between 20% δ-ALA application and blue light exposure (see Table 14.1), an incubation period used in a large US study using ALA and blue light in 243 patients with multiple AKs. In this study, 77% and 89% of the patients had 75% or more of the treated lesions cleared at weeks 8 and 12, respectively [16]. Using a similar treatment approach in 36 patients, Jeffes et al. demonstrated an 88% clearance rate of AKs, and showed a lack of correlation between lesional fluorescence and clearance rates [17]. Touma et al. reported that shorter incubation times of 1, 2 and 3 hours were sufficient for a photodynamic reaction to occur [6]. In that study 18 patients with four or more

Table 14.1. Protocol for topical PDT in epithelial cancers and precancerous lesions with different photosensitizers

Photosensitizer	ALA 20% in custom made oil-in-water emulsions or gels	Levulan Kerastick 20% solution	Metvix/Metvixia 16% cream
Application	Occlusive, light impermeable	Occlusive, light impermeable	Occlusive, light impermeable
Incubation time (h)	4–6	14–18	3–4
Light source	Blue, green or red light (indication-dependent)	Blue light (400–450 nm)	Red light from LED source
Irradiation parameters			
Light intensity (mW/cm^2)	100–180	10	Approximately 60
Light dose (J/cm^2)	120–180	10	37
Indication	Epithelial precancerous lesions, superficial basal cell carcinoma	Actinic keratoses	Superficial and nodal basal cell carcinoma and actinic keratoses
Treatments	One (retreatment if needed)	One (retreatment if needed)	Two sessions 7 days apart

AKs had their whole face treated. Thin AKs were reduced by 89.5% at 1 month and these results were maintained at 5 months. There were no differences among the 1-, 2- or 3-hour incubation groups. Szeimies and Pariser applied MAL to AKs for 3 hours followed by 75 J/cm^2 red light illumination and found a 75– 89% resolution of thin lesions, respectively [18, 19]. However, in the trial by Szeimies et al., MAL-PDT was compared to cryotherapy and showed no statistically significant difference [18]. In contrast to the current recommendations, MAL-PDT was only performed once in this trial, whereas a similar study performed in Australia revealed a significantly higher complete remission rate for MAL-PDT (91%) in the treatment of AKs, compared to a single freeze–thaw cycle of cryotherapy (68%) or placebo (30%) [20].

In a study by Ruiz-Rodriguez et al. using 20% ALA to AKs with a 4-hour incubation time followed by full-face IPL treatment with a 615-nm cut-off filter in 17 patients with photodamage [21], there was 91% resolution of AKs after two treatment sessions, with a follow-up of 3 months. Alexiades-Armenakas and Geronemus [22] used the long pulse 595-nm pulse dye laser with fluences of 4–17.5 J/cm^2, after 20% ALA incubation times of 3 or 14–18 hours, to treat patients with AKs of the face and extremities. AK clearance rates were in the 75–90% range at 3 months

with no effect of incubation times on clinical response. Topical PDT has been shown to be effective in transplant patients with diffuse AKs [23], as well as in patients with oral leukoplakia [24]. In addition, studies in UV-irradiated mice treated weekly with PDT demonstrated considerable protection from skin cancer, suggesting a potential prophylactic role in photodamaged patients with a high risk of skin cancer [25].

Patients undergoing topical PDT for AKs develop edema, erythema, crusting of AKs and generally mild to moderate discomfort in the first 24–48 hours after PDT. Recovery with desquamation of the treatment areas occurs typically within 7–10 days, depending on the severity of the underlying actinic damage. When compared to other treatment modalities such as liquid nitrogen, 5-FU, imiquimod and medium depth chemical peels, topical PDT offers comparable, if not better, clearance rates of nonhypertrophic AKs [26, 27] while allowing physician-controlled treatment, improved tolerance, and a short recovery time.

In summary, studies have shown that topical PDT clears up to 90% of nonhypertrophic AK. The advantages of PDT over other traditional therapies are its controlled nature, short recovery time, and excellent cosmetic results. Topical PDT is most useful in patients with diffuse AKs.

14

14.6 Topical PDT for Skin Rejuvenation

One of the main advantages of topical PDT is the excellent cosmetic result seen after the treated lesions have healed, unlike destructive methods that can leave hypopigmentation and scarring. In a study by Goldman, blue light PDT was used to treat 32 patients with incubation times of 15–20 hours [28]. Improvement in skin texture was seen in 72% of treated subjects. In the study by Touma et al. in 18 patients (11 women, 7 men) with AKs and mild to moderate diffuse facial photodamage, using incubation times of 1–3 hours and blue light activation of ALA, there was a statistically significant improvement in overall skin quality, notably sallowness and fine wrinkling. There was limited but noticeable improvement in mottled hyperpigmentation, but this was of borderline statistical significance. Coarse wrinkling seemed to improve in some patients but without reaching statistical significance. Patient satisfaction with the treatment was high. Patients with more severe photodamage appeared to react more intensely to PDT [6].

In the study by Ruiz-Rodriguez et al. using full-face IPL PDT with a 615-nm cut-off filter, and two sessions at an interval of 1 month [21], cosmetic results were excellent, with no pigmentary abnormalities or scarring. Broad facial PDT using IPL and pulsed dye lasers is currently being explored for the purpose of enhanced photorejuvenation. Using ALA to augment IPL and laser treatment of photodamage and vascular lesions is a promising new application, and is also currently being studied. This new indication holds the potential for reducing the need for multiple treatment sessions with these modalities. Microdermabrasion or other methods that ablate the stratum corneum may prove useful if used prior to topical PDT, as they may promote an improved and more rapid penetration of ALA.

Histology of tape-stripped skin treated with ALA PDT and a broadband light activator shows ultrastructural changes limited to the epidermis, with apoptotic damage to keratinocytes within one day of PDT, and relative sparing of melanocytes and Langerhans cells [29, 30]. This explains the improvement in skin appearance seen with topical PDT, and the limited improvement seen in the pigmentary changes of photodamage. The effect of PDT on dermal collagen is still not clear, as some studies have shown neocollagenesis [31], while others have shown that topical PDT induces the collagen-degrading metalloproteinases 1 and 3 together with a reduction in collagen production leading to an antisclerotic effect in patients with scleroderma [32]. Until further studies elucidate PDT's dermal effects, clinical experience appears to support a stimulatory role of collagen in normal skin. An unpublished limited study by Touma comparing topical ALA-PDT with trichloroacetic acid chemical peeling of diffusely actinically damaged skin has shown similar, if not improved, results, as well as neo-collagenesis, with PDT.

In summary, blue light topical PDT penetrates well enough into actinically damaged but otherwise normal skin to improve the texture and fine lines as well as the general appearance of the skin through epidermal renewal. IPL and lasers may be more beneficial for skin rejuvenation because of their depth of penetration, and their ability to treat associated signs of actinic damage, such as lentigines and telangiectasia.

14.7 Fluorescence Diagnosis

Due to the specific accumulation of the ALA-induced porphyrins in tumor cells following either topical or systemic administration, light-induced fluorescence makes the tumor cells visible. Following activation by light of the appropriate wavelength the photosensitizer molecules are excited to a higher energy state and emit during decay fluorescence. Since ROS should not be induced in fluorescence diagnosis significantly lower light intensities ($2-5$ mW/cm^2) are used as compared to PDT.

Using an optical detection system, e.g. Dyaderm Professional (Biocam, Germany), the induced fluorescence is displayed on a screen under ambient light and the localization and extension of skin tumors can be determined. For fluorescence diagnosis the strong absorption of porphyrins around 400 nm (Soret band) is used,

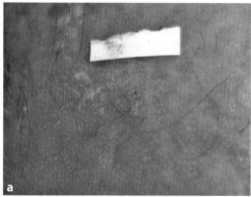

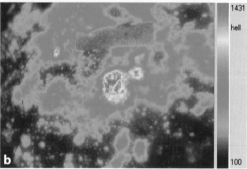

Fig. 14.4. Clinical image (**a**) and false-color coded image (**b**) of an actinic keratosis on the forehead of a 68-year-old-male using fluorescence diagnosis for better delineation. The highest fluorescence intensity (yellow–red color) indicates the margins of the lesion

and thus blue light activates the porphyrins. Because the penetration of blue light is limited to 1 mm, only superficial lesions can be detected, but the method is used very successfully intraoperatively [33].

Fluorescence diagnosis is of particular help in pretreated areas with scars and erythema, where even an experienced dermatologist has difficulty in distinguishing scar from precancerous lesions or skin cancer. Fluorescence diagnosis is an investigator-independent procedure that enables the most appropriate biopsy site to be chosen. Using digital image analysis and reference algorithms the suspected area can be shown in false colors to give maximal contrast between tumor and surrounding tissue (Fig. 14.4). The probability of false-negative biopsies is thus reduced significantly [34].

Another option is fluorescence-guided resection of a skin tumor which minimizes the tissue defect in particular in the skin, and also reduces the number of re-excisions. Image analysis enables a fluorescence threshold to be determined which defines the tumor margins. Thus, the surgeon can control the resection margins intraoperatively [35].

References

1. Szeimies RM, Dräger J, Abels C, Landthaler M (2001) History of photodynamic therapy in dermatology. In: Calzavara-Pinton PG, Szeimies RM, Ortel B (eds) Photodynamic therapy and fluorescence diagnosis in dermatology. Elsevier, Amsterdam, pp 3–16
2. Zeitouni NC, Oseroff AR, Shieh S (2003) Photodynamic therapy for nonmelanoma skin cancers. Mol Immunol 39:1133–1136
3. Marmur ES, Schmults CD, Goldberg DJ (2004) A review of laser and photodynamic therapy for the treatment of nonmelanoma skin cancer. Dermatol Surg 30:264–271
4. Ibbotson SH (2002) Topical 5-aminolaevulinic acid photodynamic therapy for the treatment of skin conditions other than non-melanoma skin cancer. Br J Dermatol 146:178–188
5. Karrer S, Abels C, Landthaler M, Szeimies RM (2000) Topical photodynamic therapy for localized scleroderma. Acta Derm Venereol 80:26–27
6. Touma D, Yaar M, Whitehead S, Konnikov N, Gilchrest BA (2004) Short incubation δ-ALA photodynamic therapy for treatment of actinic keratoses and facial photodamage. Arch Dermatol 140:33–40
7. Szeimies RM, Karrer S, Abels C, Landthaler M, Elmets CA (2001) Photodynamic therapy in dermatology. In: Krutmann J, Hönigsmann H, Elmets CA, Bergstresser PR (eds) Dermatological phototherapy and photodiagnostic methods. Springer, Berlin, pp 209–247
8. Schweitzer VG (2001) Photofrin-mediated photodynamic therapy for treatment of aggressive head and neck nonmelanomatous skin tumors in elderly patients. Laryngoscope 111:1091–1098
9. Morton CA, Brown SB, Collins S, Ibbotson S, Jenkinson H, Kurwa H, Langmack K, McKenna K, Moseley H, Pearse AD, Stringer M, Taylor DK, Wong G, Rhodes LR (2002) Guidelines for topical photodynamic therapy: report of a workshop of the British Photodermatology Group. Br J Dermatol 146:552–567
10. Liu Y, Viau G, Bissonnette R (2004) Multiple large-surface photodynamic therapy sessions with topi-

cal or systemic aminolevulinic acid and blue light in UV-exposed hairless mice. J Cutan Med Surg 8:131–139

11. Stender IM, Bech-Thomsen N, Poulsen T, Wulf HC (1997) Photodynamic therapy with topical delta-aminolevulinic acid delays UV photocarcinogenesis in hairless mice. Photochem Photobiol 66:493–496

12. Henderson BW, Dougherty TJ (1992) How does photodynamic therapy work? Photochem Photobiol 55:145–157

13. Karrer S, Bäumler W, Abels C, Hohenleutner U, Landthaler M, Szeimies RM (1999) Long-pulse dye laser for photodynamic therapy: investigations in vitro and in vivo. Lasers Surg Med 25:51–59

14. Food and Drug Administration (1999) First Drug Device Combined Treatment for Certain Pre-cancerous Skin Lesions Approved. US Department of Health and Human Services. http://www.fda.gov/bbs/topics/NEWS/NEW00704.html

15. Food and Drug Administration (2003) Levulan Kerastick (aminolevulinic acid HCl) for Topical solution, 20%. http://www.fda.gov/medwatch/SAFETY/2003/03MAR_PI/Levulan_Kerastick_PI.pdf

16. Piacquadio DJ, Chen DM, Farber HF, Fowler JF Jr, Glazer SD, Goodman JJ, Hruza LL, Jeffes EW, Ling MR, Phillips TJ, Rallis TM, Scher RK, Taylor CR, Weinstein GD (2004) Photodynamic therapy with aminolevulinic acid topical solution and visible blue light in the treatment of multiple actinic keratoses of the face and scalp: investigator-blinded, phase 3, multicenter trials. Arch Dermatol 140:116–120

17. Jeffes EW, McCullough JL, Weinstein GD, Kaplan R, Glazer SD, Taylor JR (2001) Photodynamic therapy of actinic keratoses with topical aminolevulinic acid hydrochloride and fluorescent blue light. J Am Acad Dermatol 45:96–104

18. Szeimies RM, Karrer S, Radakovic-Fijan S, Tanew A, Calzavara-Pinton PG, Zane C, Sidoroff A, Hempel M, Ulrich J, Proebstle T, Meffert H, Mulder M, Salomon D, Dittmar HC, Bauer JW, Kernland K, Braathen LR (2002) Photodynamic therapy using topical methyl 5-aminolevulinate compared with cryotherapy for actinic keratosis: a prospective, randomized study. J Am Acad Dermatol 47:258–262

19. Pariser DM, Lowe NJ, Stewart DM, Jarrat MT, Lucky AW, Pariser RD, Yamauchi PS (2003) Photodynamic therapy with topical methyl aminolevulate for actinic keratosis: results of a prospective randomized multicenter trial. J Am Acad Dermatol 48:227–232

20. Freeman M, Vinciullo C, Francis D, Spelman L, Nguyen R, Fergin P, Thai KE, Murrell D, Weightman W, Anderson C, Reid C, Watson A, Foley P (2003) A comparison of photodynamic therapy using topical methyl aminolevulinate (Metvix®) with single cycle cryotherapy in patients with actinic

keratosis: a prospective, randomized study. J Dermatol Treat 14:99–106

21. Ruiz-Rodriguez R, Sanz-Sanchez T, Cordoba S (2002) Photodynamic photorejuvenation. Dermatol Surg 28:742–744

22. Alexiades-Armenakas M, Geronemus RG (2003) Laser-assisted photodynamic therapy of actinic keratoses. Arch Dermatol 139:1313–1320

23. Dragieva G, Hafner J, Dummer R, Schmid-Grendelmeier P, Roos M, Prinz BM, Burg G, Binswanger U, Kempf W (2004) Topical photodynamic therapy in the treatment of actinic keratoses and Bowen's disease in transplant recipients. Transplantation 15:115–121

24. Sieron A, Adamek M, Kawczyk-Krupka A, Mazur S, Ilewicz L (2003) Photodynamic therapy (PDT) using topically applied delta-aminolevulinic acid (ALA) for the treatment of oral leukoplakia. J Oral Pathol Med 32:330–336

25. Sharfaei S, Juzenas P, Moan J, Bissonnette R (2002) Weekly topical application of methyl aminolevulinate followed by light exposure delays the appearance of UV-induced skin tumours in mice. Arch Dermatol Res 94:237–242

26. Kurwa HA, Yong-Gee SA, Seed PT, Markey AC, Barlow RJ (1999) A randomized paired comparison of photodynamic therapy and topical 5- fluorouracil in the treatment of actinic keratoses. J Am Acad Dermatol 41:414–418

27. Dinehart SM (2000) The treatment of actinic keratoses. J Am Acad Dermatol 42:25–28

28. Goldman MP, Atkin D, Kincaid S (2002) PDT/ALA in the treatment of actinic damage: real world experience. Laser Surg Med [Suppl] 14:79

29. Bartosik J, Stender IM, Kobayasi T, Agren MS (2004) Ultrastructural alteration of tape-stripped normal human skin after photodynamic therapy. Eur J Dermatol 14:91–95

30. Nakaseko H, Kobayashi M, Akita Y, Tamada Y, Matsumoto Y (2003) Histological changes and involvement of apoptosis after photodynamic therapy for actinic keratoses. Br J Dermatol 148:122–127

31. Fink-Puches R, Soyer HP, Hofer A, Kerl H, Wolf P (1998) Long-term follow-up and histological changes of superficial nonmelanoma skin cancers treated with topical delta-aminolevulinic acid photodynamic therapy. Arch Dermatol 134:821–826

32. Karrer S, Bosserhoff AK, Weiderer P, Landthaler M, Szeimies RM (2003) Influence of 5-aminolevulinic acid and red light on collagen metabolism of human dermal fibroblasts. J Invest Dermatol 120:325–331

33. Abels C (2001) Fluorescence diagnosis. In: Calzavara Pinton G, Szeimies RM, Ortel B (eds) Photodynamic therapy and fluorescence diagnosis in dermatology. Elsevier, Amsterdam, pp 165–176

34. Ericson MB, Sandberg C, Gudmundson F, Rosen A, Larkö O, Wennberg AM (2003) Fluorescence contrast and threshold limit: implications for photodynamic diagnosis of basal cell carcinoma. J Photochem Photobiol B 69:121–127

35. Stummer W, Reulen HJ, Novotny A, Stepp H, Tonn JC (2003) Fluorescence-guided resections of malignant gliomas – an overview. Acta Neurochir Suppl 88:9–12

14

Chapter 15

Botulinum Toxin, Fillers, Peels: The Scientific View

15

Nanna Schürer

Contents

15.1 Introduction

According to the American Society of Plastic Surgeons (ASPS), people continue to invest in their looks. Botulinum toxin injections increased by 2,400% from 1997 to 2003. Of all "rejuvenation procedures" surgery constitutes 25% and noninvasive techniques 75%. Nearly 7 million procedures were performed in 2002 in the US, but the underlying hopes and promises are not always based on accurate analysis. This review focuses on the scientific basis of relatively noninvasive techniques, such as botulinum toxin injections, fillers and peels. Techniques and procedures are discussed elsewhere.

15.2 Botulinum Toxin

Botulinum toxin (BTX-A) is used to treat facial lines [29]. Carruthers and Carruthers were the first to report the use of BTX-A for the improvement of the glabellar frown in 1992 [5]. Used originally to improve rhytides caused by hyperkinetic movement and muscular hypertrophy in the upper one-third of the face, aesthetic indications including common off-label location use continue to expand. BTX has been approved by the US Food and Drug Administration (FDA) for neuromuscular disorders (cervical dystonia, hemifacial spasm, blepharospasm, strabismus). Recently BTX-A has also been accepted by the FDA and the Canadian Health Protective Branch for the treatment of glabellar frown lines [32].

Table 15.1. Target substrates of BTX (*SNAP* synaptosomal associated protein, *VAMP* vesicle associated membrane protein)

Botulinum toxin type	Substrate
A	SNAP-25
B	VAMP (synaptobrevin)
C	SNAP-25 (syntaxin)
D	VAMP (synaptobrevin)
E	SNAP-25
F	VAMP (synaptobrevin)
G	VAMP (synaptobrevin)

Table 15.2. Toxins commercially available in the USA and Europe

Type	Trade name (US)	Trade name (Europe)	Units/ vial	Form	Company	Stability (months)
A	Botox	Botox	100	Freeze-dried crystals	Allergan	12
A		Dysport	500	Freeze-dried crystals	Ipsen Biopharm	12
B	Myobloc	Neurobloc	2,500	Solution	Elan Pharmaceuticals	12

15.2.1 Types of Toxins

Neurotoxins produced by *Clostridium botulinum* are the most potent toxins known to man and are the causative agents of botulism. *Clostridium botulinum* is a gram-positive, anaerobic bacterium that is widely distributed in the environment, mainly as spores. There are eight structurally similar but antigenically different strains produced by *C. botulinum*, and the human nervous system is susceptible to only five toxin subtypes (A, B, E, F and G) and is unaffected by three (C-1, C-2 and D). All BTX have the same basic mechanism for producing their paralytic effect. They cause a chemical denervation resulting in a temporary paralysis of exposed striated muscles. Injection of toxin into facial muscles weakens the contractions of the exposed muscles and the corresponding pull of the superimposed skin, thereby reducing wrinkles and furrows. The antagonizing muscle efficacy increases and therefore a modulation of facial expression is notable.

Neurotoxicity is accomplished through a sequence of three events following toxin exposure of striated muscle:

■ Neurotoxin binding to specific membrane receptors on presynaptic cholinergic neurons

■ Toxin/receptor complex internalization via endocytosis into nerve terminals

■ Vesicle lysis and prevention of acetylcholine release from inside the cell

Each of the toxins binds to a separate target protein on the cell membrane and has a separate target protein inside the cell (Table 15.1). This accounts for the lack of cross-reactivity between toxin serotypes and contributes to both clinical and immunological disparities. If neutralizing antibodies and immunoresistance occur to one form, rendering the patient unresponsive to

Table 15.3. Characteristics of BTX types

Trade name (Europe)	pH	Complex size (kDa)	Storage/stability (months)
Botox	7.3	900	12 (crystallized)
Dysport	6.0-7.0	500	12 (crystallized)
Xeomin		150	36 at room temperature (crystallized)
Neurobloc	5.6	700	12 (liquid)

treatment, an alternative may be used to produce the same effect.

There are three commercially available toxins. Type A is considered the most toxic and comes in two forms: in the US as Botox (Allergan) and in Europe also as Dysport (Ipsen Biopharm) (Table 15.2). Botox is produced by a multiple precipitation technique and is distributed in vials containing 100 U lyophilized crystal. Dysport is the result of column-based purification and is provided in vials containing 500 U. Although both forms of type A toxin require the addition of a diluent before use, the recommended doses of the two are markedly different. Immunoresistance to Botox is not adequately documented when using small doses. However, blocking antibodies resulting in treatment failure have appeared when high doses (more than 400 U) of Botox were administered to patients with cervical dystonia. This prompted investigation into the use of an alternate toxin.

BTX-B is distributed by Elan Pharmaceuticals as Myobloc in the US and Neurobloc in Europe (Table 15.2). Most experience with BTX-A has been for the treatment of cervical dystonia. Myobloc comes ready to use as a liquid formulation in vials of 2,500 U/0.5 ml, 5,000 U/1.0 ml and 10,000 U/2.0 ml. Little information is available regarding dosing equivalents between toxin subtypes A and B. For the treatment of cervical dystonia 5,000–10,000 U of Myobloc is therapeutic, whereas the equivalent dose of Botox is 100–300 U, suggesting that Botox is 50 to 100 times more effective than Myobloc (Table 15.3).

15.2.2 Studies

Most studies from the US have concentrated on the use of Botox for the treatment of facial expression lines, but there are a few studies in which Dysport has been employed for that indication. The potencies of Botox and Dysport in the treatment of cervical dystonia have been compared in a double-blind randomized crossover study [36]. Each patient received three treatments (first Botox, second Dysport 1:3, and third Dysport 1:4), and each treatment period lasted 16 months. Employing a crossover design, each patient acted as his or her own control. Dysport had a better effect on impairment and pain than Botox with a conversion factor of 3.

The duration of muscle paralysis and expression line reduction following treatment with BTX-A and BTX-B have been compared in only a few double-blind studies. Matarasso evaluated the effect of BTX-A in comparison with that of BTX-B for the treatment of canthal rhytides in ten women [32]. For this double-blind randomized evaluation the lowest effective ratio of 1:50 (1 U of BTX-A to 50 U of BTX-B) was chosen. The average onset of the effect of BTX-B was one day earlier than that of BTX-A, discomfort on injection was greater with BTX-B than with BTX-A, muscle paralysis on day 7 after injection was comparable for both BTX-A and BTX-B, and the average duration of the effect of BTX-B was 6 weeks while that of BTX-A was 12.7 weeks.

Reduction of injection pain with BTX-B was studied in 15 patients receiving a total of 5,250 U BTX-B on both sides of the face, diluted with saline containing either no preservative or benzyl alcohol as preservative [41]. The BTX-A:BTX-B ratio in this study was 1:125. Out of 15 patients,

13 found the injections of the preservative-containing solution less painful. Benzyl alcohol is known to have an anaesthetic effect. However, the possibility of late-type sensitization has not been considered so far.

A prospective follow-up for 1 to 3 years was performed in 52 patients treated with BTX-A for facial rhytides [4]. Repeat injections were required every 3 to 6 months to maintain the desired improvement. The muscles of the frown, forehead and eyes are intricately intertwined. All of the superficial muscles of the face are part of the superficial musculoaponeurotic system. Recognizing their relationships allows the desired cosmetic result to be achieved and compensation and adverse events to be anticipated [42].

15.2.2.1 Dose-Finding Studies

Carruthers et al. compared the safety and efficacy of three doses of Botox for the treatment of horizontal forehead rhytides in a controlled setting [6]. Of 59 patients enrolled, 20 received the total dose of 16 U BTX-A, 19 received 32 U and 20 received 48 U. Patients were evaluated at regular intervals up to 48 weeks after injection by trained observers and self-assessment. Improvement of forehead rhytides was greater with 48 U BTX-A than with 16 U BTX-A, and the time until relapse was also longer.

Flynn et al. performed another dose-finding study of BTX-A for the treatment of lower eyelid rhytides [19]. The lower eyelids of 19 women were injected bilaterally with either 4 or 8 U BTX-A (Botox). Unilaterally, 12 U BTX-A was injected into the crow's feet area 1 cm outside the orbital rim. With increasing doses of BTX-A a dose-response curve was seen, but the higher dose, i.e. 8 U, gave unattractive results. For eye widening, the authors advise the use of a maximum of 4 U in the lower eyelid. For the treatment of periocular wrinkles 2 U in the lower eyelid along with 12 U in the crow's feet area is recommended.

15.2.2.2 Double-Blind Placebo-Controlled Studies

In the first double-blind placebo-controlled randomized multicenter clinical study of the efficacy and safety of BTX-A (Botox) in the treatment of glabellar lines [8], 264 patients were enrolled, of whom 203 were assigned to BTX-A treatment and 61 to the placebo treatment. A total dose of 20 U BTX-A in a total volume of 0.5 ml, or normal saline as placebo, was injected. Evaluation was performed by trained observers and patient self-assessment at regular intervals up to 120 days after injection. BTX-A was superior to placebo. According to the patients' self-assessment, there was no change after placebo treatment, while a 75% improvement was noted in the BTX-A group at 30 and 60 days after injection. However, the trained observers rated glabellar line severity at rest to be 23–29% improved in the placebo group 30 and 60 days after injection, although the response rate in the BTX-A group was significantly higher.

Another placebo-controlled bilateral double-blind randomized comparison of BTX-A was performed in 60 patients [29]. Patients were treated with 6, 12 or 18 U BTX-A (Botox) into the orbicularis muscle on one side and placebo (normal saline) on the other side. Evaluation was performed by trained observers and patient self-assessment at regular intervals up to 16 weeks after injection. All doses of BTX-A were superior to placebo with no clear dose-response relationship. Interestingly, the patients' self-assessment revealed a placebo success rate of between 10% and nearly 50%. However, the trained observers assessed the placebo success rate as 18% at most.

A third double-blind placebo-controlled study on the efficacy and safety of BTX-A (Botox) was performed on patients with glabellar lines of at least moderate severity [9]. Of 273 patients enrolled, 202 received BTX-A treatment and 71 placebo treatment. Again, a total dose of 20 U BTX-A in a total injection volume of 0.5 ml, or normal saline as placebo, was injected. Evaluation was performed by trained observers and patient self-assessment at regular intervals up to 120 days after injection. Again, BTX-A was superior to placebo. According to the patients' self-assessment, the response rate in the BTX-A group peaked at 88% and that in the placebo group at 13%. According to the trained observer assessment at maximum frown, the glabellar line severity values of the placebo group remained near baseline values throughout the study. However, the at-rest responder analysis revealed a response rate in the BTX-A group peaking at 70%, while that in the placebo group remained below 35% throughout the study.

15.2.3 Scientific View

The use of BTX-A for the treatment of facial expression lines has been shown to be effective in several double-blind placebo-controlled randomized studies. Further, the FDA has approved BTX-A for the treatment of glabellar frown lines. The mechanisms of action of several toxin subtypes have been studied. Therefore, there is no doubt that the use of BTX-A is an effective tool – as long as the facial anatomy and muscular tonus are understood.

15.3 Fillers

Injectable materials suitable for soft-tissue augmentation and volume expansion have been studied for more than 40 years. The ideal injectable material for filling wrinkles and restoring volume should be easy to use and offer long-lasting aesthetically pleasing results, with a minimal risk of unwanted side effects. Fillers include synthetic products (silicone, paraffin, acrylates) and natural substances (collagen, autologous fat, hyaluronic acid). Here, resorbable (temporary) and nonresorbable (permanent) fillers are discussed.

15.3.1 Temporary Fillers

Details of temporary filler products including trade names and other product information are provided in Table 15.4.

15.3.1.1 Hyaluronic Acid

Hyaluronic acid (HA) is a complex sugar that is composed of repeating subunits of D-glucuronic acid and N-acetyl-glucosamine in long unbranched polyanionic chains. In its non-crosslinked form, HA has a tissue half-life of 1–2 days. Crosslinking between the long polysaccharide chains forms a hydrophilic insoluble polymer and dramatically increases the tissue residence time.

HA-containing injectable products are approved by the FDA for treating wrinkles. FDA approval was based on a review of the clinical studies conducted by the manufacturer and on the recommendation of the General and Plastic Surgery Devices Panel of the FDA Medical Devices Advisory Committee.

Chemical Evaluation

A chemical comparative evaluation was performed by Manna and coworkers [31], comparing Restylane with Hylaform (Table 15.5). Hylaform Viscoelastic Gel (Hylan B) is derived from rooster combs and subjected to crosslinking, and Restylane is produced through bacterial fermentation (Streptococci) and stabilized. The amount of crosslinked HA in Hylaform is about three-quarters that in Restylane. Restylane contains protein resulting from bacterial fermentation or added protein to enable the crosslinking reaction. The quantity of proteins contained by Restylane can be as much as four times the quantity contained by Hylaform [31]. Hylaform may cause an allergic reaction in people sensitized to avian products (Table 15.4).

High molecular weight (6×10^6 Da) molecules may initiate foreign body granulomatous reactions after intradermal injection [30]. The biocompatibility of Hylan G-F 20 (Synvisc, Genzyme Biosurgery, Ridgefield, N.J.) has been compared with that of saline and unmodified HA 28 days after injection in guinea pigs [37], producing both cell-mediated and specific humoral reactions. Hylan was a strong sensitizer in guinea pigs because of the specific antibody production induced after one injection. However, unmodified HA had no immunogenicity in experimental animals.

15

Table 15.4. Temporary injectable fillers, including product information, possible side effects and regulatory status

Filler	Company	Product information	Purpose	Side effects	Results	Regulatory status
Bovine-derived collagen						
Zyderm I and II	Inamed Aesthetics, Palo Alto, Calif.	From purified cows' skin	Fills wrinkles, lines and scars on face and around lips	Allergic reaction; requires skin test prior to procedure.	Immediate; lasts up to 6 months	FDA-approved
Zyplast	Inamed Aesthetics, Palo Alto, Calif.	From purified cows' skin	Fills deep wrinkles, pronounced lines and facial scars	Allergic reaction; requires skin test prior to procedure.	Immediate; lasts up to 6 months	FDA-approved
Human-derived collagen						
Autologen	Collagenesis, Beverly, Mass.	From the patient's skin	An alternative to traditional collagen injections	Bruising; time-consuming and expensive	Not permanent	Not required
Cymetra (micronized Alloderm)	LifeCell, Branchburg, N.J.	From cadaver tissue, screened for contamination	For lips, nasolabial folds, deep wrinkles	Bruising; does not need skin test	Multiple treatments needed	FDA-approved
CosmoDerm	Inamed Aesthetics, Palo Alto, Calif.	From human tissue that has been purified and grown in a laboratory	For superficial lines	Bruising; does not need skin test	Immediate; lasts up to 6 months	FDA-approved
CosmoPlast	Inamed Aesthetics, Palo Alto, Calif.	From human tissue that has been purified and grown in a laboratory	For pronounced wrinkles	Bruising; does not need skin test	Immediate; lasts up to 6 months	FDA-approved
Dermalogen	Collagenesis, Beverly, Mass.	From cadaver tissue, screened for contamination	An alternative to traditional collagen	Erythema and acneiform rash	Three treatments needed	FDA-approved
Fascian	Fascia Biosystems, Beverly Hills, Calif.	From donor fascia (connective tissue made of collagen) of the thigh muscle	Stimulates collagen formation	Bruising	Lasts up to 6 months	FDA-approved
Isolagen	Isolagen Laboratories, Houston, Tx.	Autologous fibroblast cultures	For superficial lines	Time-consuming and expensive	Lasts up to 12 months	Not required

Table 15.4. *Continued*

Filler	Company	Product information	Purpose	Side effects	Results	Regulatory status
Human-derived product						
Plasmagel, Vitagel		Plasma emulsion (protein) made of patient's blood and vitamin C complex	Soft-tissue filler to add volume.	Bruising; does not need skin test	Lasts up to 3 months	Not required
Hyaluronic acid						
Restylane	Q-Med, Sweden	Bacterial fermentation non-animal stabilized HA	Medium density product	Redness, swelling, tenderness	Immediate; lasts 6 months	FDA-approved for filling wrinkles
Restylane Fine-lines	Q-Med, Sweden	Bacterial fermentation non-animal stabilized HA	Low-density product	Redness, swelling, tenderness	Immediate; lasts 6 months	In use in Europe
Perlane	Q-Med, Sweden	Bacterial fermentation non-animal stabilized HA	High-density product	Redness, swelling, tenderness	Immediate; lasts 6 months	In use in Europe
Macrolane	Q-Med, Sweden	Nonanimal stabilized HA	Very high-density product	Redness, swelling, tenderness		In use outside US
Rofilan Hylan Gel	Q-Med, Sweden	Nonanimal stabilized HA	Low-density product	Redness, swelling, tenderness		In use outside US
Hyal System	Merz Pharmaceuticals	From rooster combs; no crosslinks	Soft-tissue filler to add volume		Immediate; short duration	In use outside US
Hylaform	Inamed Aesthetics	From rooster combs	Medium-density product	Redness, swelling; people with sensitivities to avian products may have an allergic reaction	Varies; lasts up to 6 months	FDA-approved for filling wrinkles
Hylaform Fineline	Inamed Aesthetics	From rooster combs; cross-links with vinylsulfon	Low-density product	Redness, swelling; people with sensitivities to avian products may have an allergic reaction		In use outside US

Table 15.4. *Continued*

Filler	Company	Product information	Purpose	Side effects	Results	Regulatory status
Hylaform Plus	Inamed Aesthetics	From rooster combs; cross-links with vinylsulfon	High-density product	Redness, swelling; people with sensitivities to avian products may have an allergic reaction		FDA-approved for filling wrinkles
Viscontour	Aventis Germany	Bacterial fermentation		Redness, swelling, tenderness		In use outside US
Juvederm 18	Euromedical Systems Limited, UK	Nonanimal stabilized HA	Low-density product	Redness, swelling, tenderness		In use outside US
Juvederm 24	Euromedical Systems, UK	Nonanimal stabilized HA	Medium-density product	Redness, swelling, tenderness		In use outside US
Juvederm 30	Euromedical Systems, UK	Nonanimal stabilized HA	High-density product	Redness, swelling, tenderness		In use outside US
Juvelift	Corneal, France	Nonanimal stabilized HA	Soft-tissue filler that adds volume	Redness, swelling, tenderness, bruising	Immediate; lasts 6 months	In use in Europe
Matridex	BioPolymer	Nonanimal stabilized HA	50% HA, 25% dextran	Redness, swelling, tenderness, bruising	Immediate; lasts 12 months	In Europe (MDD 93/42/ EWG)
Matridur	BioPolymer	Nonanimal stabilized HA	50% HA	Redness, swelling, tenderness, bruising	Immediate; lasts 6 months	In Europe (MDD 93/42/ EWG)

15

Table 15.5. Chemical characteristics of two different forms of crosslinked HA (according to Ref. 31)

	Restylane	Hylaform
Source	Bacterial fermentation	Rooster combs
Hydrogel	Weak	Strong
Rheological properties		Superior to Restylane
Protein content (crosslinked HA)		25% of Restylane
Clinical experience		20 years

Clinical Studies

A clinical and histological study performed on 158 volunteers undergoing soft-tissue augmentation of the face with Restylane revealed a subjective improvement in 80% of subjects over an 8-month treatment course. Minor side effects were documented in 12% of subjects, but subsided in a few days [15]. Inflammatory local reactions to any HA-containing filler may occur 1 to 14 days after treatment with an incidence of 0.4%.

In a randomized evaluator-blinded multicenter evaluation of the safety and efficacy of Restylane in comparison with Zyplast for the correction of nasolabial folds [34], 138 volunteers received one of the treatments on each side of the face. Treatments were repeated at 2-week intervals. Restylane was found to be somewhat superior in approximately 60% of patients, whereas Zyplast was superior in approximately 10%. However, as reported by patients, within 14 days following the first treatment, the Restylane-treated side had a comparable incidence of severe redness (5.1% vs 5.8%) but an increased incidence of severe bruising (3.6% vs 0.7%), swelling (3.6% vs 1.4%), pain (3.6% vs 1.4%), and tenderness (2.9% vs 1.4%) compared with the Zyplast-treated side. These incidences were lower following follow-up injections for both products.

In a prospective, randomized study analysing the effect of BTX-A and Restylane compared to Restylane alone in parallel groups of 19 patients with deep resting glabellar lines, the combined treatments provided superior aesthetic benefit and longevity of response versus Restylane alone [7].

There are limited safety data for Restylane in non-Caucasians, although the manufacturer, Q-Med AB of Sweden, has agreed to conduct a post-approval study in people of colour to determine the product's safety for this population. The company will also provide training to physicians on the correct use of the device. The rates of side effects reported after injection of Restylane and Perlane were 1.5% in 1999 (1,444,000 patients) and 0.6% in 2000 (262,000 patients) [20].

A study in 20 volunteers employing the non-crosslinked HA Hyal System for soft-tissue augmentation of the face revealed an increased skin turgor as measured using an EM25 skin elastometer [13]. However, a double-blind multicenter comparison has been done only for Restylane. In summary, the crosslinked form of HA, either derived from rooster combs or bacterial fermentation, has been used for soft-tissue augmentation for several years. The risk of side effects, including discoloration, oedema, erythema and induration, is minimal for this temporary filler.

15.3.1.2 Collagen

Since the 1980s, injectable collagen has been used as a soft-tissue filler. Collagen is a naturally occurring protein that supports various parts of the body including skin, tendons and ligaments. A collagen molecule consists of a triple helix of large peptide chains (300 nm long), i.e. glycine, lysine and proline. According to the ASPS, in 2003 more than 576,000 people had collagen injections, supporting the assertion that collagen is the most widely used filler material.

Human Collagen

A number of companies offer human collagen, either obtained from the donor such as Autologon, introduced in the 1980s (Collagenesis, Beverly, Mass.) or obtained from cadavers, such as Alloderm/Cymetra (LifeCell, Branchburg, N.J.) or Dermatologon (Collagenesis). These products are more widely used in the US than in Europe. Preparation of these fillers is expensive and time consuming, but significant side effects have not been reported. Still, bruising is common and, with one product, prolonged erythema and acneiform rashes have been noted in 10% of 130 treated patients [16, 17].

Bovine Collagen

Commonly used injectable collagen is made from purified cows' skin, and is used to fill wrinkles, lines and scars on the face. It is absorbed into the body. Bovine collagen implants are temporary fillers and their use has been associated with major adverse side effects. Approximately 3% of patients experience an allergic reaction to injected bovine collagen with development of circulating antibodies to the foreign material. This reaction subsides within 6 months, but has been reported to last up to 18 months in some cases. Double skin testing is therefore recommended, but not required, prior to treatment with collagen to monitor for this allergic response [16]. However, it has to be born in mind that sensitization may be initiated upon any collagen injection. Therefore, skin testing is no guarantee of compatibility.

Although collagen is the most frequently used filler, the only recent evaluator-blinded comparison of safety and efficacy is the above-mentioned study by Narins et al. [34].

15.3.2 Alloplastic Materials

For more than 20 years microsphere-encapsulated synthetic polylactic acid (PLA), a type of aliphatic polyester has been used in a number of therapeutic contexts. PLA occurs in the L-, D-, *meso*- and racemic forms. L-PLA has a molecular weight of >100,000 Da and is resorbed within 12 to 24 months. The synthetic material is degraded in vivo into lactic acid (LA), which the body metabolizes in the lactic cycle to CO_2. Sculptra contains 150 mg PLA, sodium caramellose 90 mg and mannit 127.5 mg, and after subdermal injection has two effects: (1) the gel itself provides benefits, and (2) with time neo-collagenesis is observed.

PLA has been approved by the FDA for the treatment of facial lipoatrophy based on a study of 30 HIV-positive individuals with facial lipoatrophy (based on physician assessment) given PLA in three bilateral injections 2 weeks apart into the deep dermis overlying the buccal fat pad [33]. Assessments included facial ultrasound, visual analogue scales, and photographs. No changes in immunological, virological, biochemical, haematological or metabolic parameters emerged during the study. Injections were well tolerated with only two adverse events (cellulitis and bruising) recorded. Patient visual analogue assessments, photographic assessments, and anxiety and depression scores at weeks 0, 12 and 24 improved during treatment.

15.3.3 Permanent Fillers

Details of temporary filler products including trade names and other product information are provided in Table 15.6.

15.3.3.1 Silicone

Injectable silicone is not approved for cosmetic use in the US. In 1991 the FDA banned its use for the treatment of wrinkles and facial defects due to its tendency to harden, migrate and cause inflammation and skin necrosis.

15.3.3.2 Polymethylmethacrylate Microspheres

A mixture of collagen and polymethylmethacrylate microspheres was developed by Lemperle and introduced in Europe in 1992. However, superficial granulomatous reactions are frequent with acrylate-containing fillers, and their use should be discouraged.

15

Table 15.6. Permanent injectable fillers, including product information, possible side effects and regulatory status

Filler	Company	Product information	Purpose	Side effects	Results	Regulatory status
Artecoll	KEMA, The Netherlands	75% bovine collagen and 25% polymethylmethacrylate microspheres (nonsilicone, carbon-based polymers)	Collagen formation around the microspheres	Granulomas (foreign body reactions), microspheres can move to other areas of the body	Immediate; 3 months between injections	FDA advisory panel recommended approval
Radiance/Radiesse	Bioform Medical, The Netherlands	Calcium hydroxyapatite (found in bone and teeth, made into an injectable paste)	Little risk of allergic reaction, collagen formation around the microspheres	Lumping, granulomas (foreign body reactions), microspheres can move to other areas of the body	Immediate; reported to last 2–5 years	FDA-approved for vocal cord and dental defects
Aquamid	Contura, Denmark	Polyacrylamide gel, inline crosslinking technology	Collagen formation around the microspheres	Granulomas (foreign body reactions)	Immediate; reported to last many years	In use in Europe
DermaLive/DermaDeep	Euromedical Systems, UK	Acrylhydrogel–hyaluronic acid	Collagen formation around the microspheres	Granulomas (foreign body reactions)	Immediate; reported to last many years	In use in Europe
Bio-Alcamid	Polymekon, Italy	Polyacrylamide gel	Collagen formation around the microspheres	Granulomas (foreign body reactions)	Immediate; reported to last many years	In use in Europe
Outline	Medical Aesthetic, France	Polyacrylamide gel	Collagen formation around the microspheres	Granulomas (foreign body reactions)	Reported to last 2–5 years	In use in Europe
Bioinblue	Polymekon, Italy	Polyvinyl alcohol	Collagen formation around the microspheres	Redness, swelling, tenderness, bruising		In use in Europe
Silikon 1000				Migrate to unwanted areas; granulomas (foreign body reactions)		FDA-approved for certain eye-related injections, but not for cosmetic corrective use

Table 15.6. *Continued*

Filler	Company	Product information	Purpose	Side effects	Results	Regulatory status
Alloplastic material						
Sculptra	Aventis	Synthetic polylactic acid contained in microspheres	Restores lost facial volume in people with HIV; collagen formation around the microspheres	Redness, bruising, lumping; deep tissue granulomas	Immediate; typically last 2 years	FDA-approved for facial lipoatrophy

15

15.3.3.3 Feather-Lift

The "Feather-Lift" is a non-surgical approach to facial lifting utilizing patented "Aptos threads". Aptos threads are made of polypropylene monofilaments, and are designed with directional cogs. Once inserted under the skin, the Aptos thread cogs form a support structure for the tissue of the face.

15.3.4 Side Effects

Histological and clinical features of granulomas due to orofacial fillers have been described in 11 patients. Only 3 of the 11 patients knew the nature of the injected product. Classic giant cell granulomas have been associated with the use of Artecoll, Dermalive, and New-Fill, and a cystic and macrophagic histological pattern has been associated with the use of liquid silicone. Information is often missing or misleading because patients and practitioners may be reluctant to give details [28].

15.3.5 Scientific View

There are many injectable fillers on the market. Some are presented in Tables 15.4 and 15.6. Whether a filler is synthetic or from "natural sources", side effects such as the generation of granulomas must be considered. Using temporary fillers these reactions are usually mild and subside as the filling effect subsides. In contrast, permanent fillers may provoke foreign body reactions that persist for months and years after injection. Bearing in mind that all of these modalities are employed for aesthetic purposes in healthy people, the incidence of possible side effects is high. Furthermore, there are only very few blinded placebo-controlled randomized studies that substantiate the current claims. An accessible international repository of efficacy and safety data of injectable fillers is needed.

Table 15.7. Classification of chemical peeling

Depth of injury	Chemicals
Superficial: epidermal	AHA, SA, Jessner
Medium: papillary to upper reticular dermis	TCA, combination peel
Deep midreticular dermis	Phenol, TCA in combination with other peeling substances

15.4 Chemical Peels

Chemical peeling is defined as the application of one or more chemical exfoliating agents to the skin, resulting in destruction of portions of the epidermis or dermis followed by regeneration of new epidermal and dermal tissue. Several chemical agents are currently used to perform chemical peels of the face. These include:

1. Alpha-hydroxy acids (AHAs, e.g. glycolic acid, GA, 20–70%)
2. Salicylic acid (SA, 30%)
3. Resorcinol
4. Pyruvic acid/pyruvate
5. Trichloroacetic acid (TCA, 15–30%)
6. Jessner's solution (14% LA, 14% resorcinol, 14% SA)
7. Combinations of agents 1–6
8. Phenol

The depth of wounding produced by chemical peeling depends on the strengths of the agent employed. Classification of chemical peeling has been defined according to the histological depth of injury exerted by the agent (Table 15.7).

15.4.1 Alpha-Hydroxy Acids

For more than 20 years, AHAs have been used in dermatology as superficial peeling agents. AHAs are organic acids with one hydroxyl group attached to the alpha position of the acid. AHAs belong to the family of "fruit acids", which are common ingredients of fruits. GA, for example, is present in grapes and cane sugar. Accordingly, the "naturally derived" claim is appealing, but in reality, AHAs of synthetic origin are more commonly used.

15.4.1.1 Alpha-Hydroxy Acids as Cosmetic Ingredients

AHAs can significantly reverse both epidermal and dermal markers of photoaging. AHAs are also used to treat epidermal lesions or those in the superficial dermis, including fine wrinkles, actinic keratoses, melasma, lentigines, and seborrhoeic keratoses. Therefore, AHAs may not be pure cosmetic ingredients. However, until now few side effects have been reported. In the US, formulations containing GA at concentrations of 1–8% are on the market without any prescription. Higher concentrations require medical supervision. In Europe no legislation exists with respect to the use of AHAs, except in Switzerland, where a maximal concentration of 10% AHA with a minimal pH of 3.5 is allowed for cosmetic use.

Low-pH formulations may irritate sensitive skin upon repeated application disturbing the physiological pH of between 4.2 and 5.6. Lower concentrations or partially neutralized buffered AHAs are present in creams and lotions. Claims for improving skin surface appearance and rejuvenation properties have led to more than 160 new cosmetic products in 1994 alone. Despite the extensive use of AHAs in cosmetic products and peeling procedures, their mechanisms of action are not well understood. Several hypotheses have been proposed.

Although rigorous proof is missing, the literature reveals some interesting studies; these are described below.

A placebo-controlled study in 17 volunteers applying a lotion containing 25% GA, LA, or citric acid (pH 3.5) to one forearm and a placebo lotion to the opposite forearm for an average of 6 months was one of the first histological and ultrastructural studies performed. The AHAs caused an increase of approximately 25% in skin thickness. The epidermis was thicker and papillary dermal changes included increased acid mucopolysaccharides, improved quality of elastic fibres, and increased density of collagen [14]. Smith [38] studied the effects on skin firmness and thickness of 5% and 12% LA test solutions (pH 2.8) applied twice daily over 16 weeks. Treatment with 12% LA resulted in significant thickening of the epidermis and dermis. Neither study provided objective clinical assessments, however.

DiNardo et al. [12] studied the effects of formulations containing various concentrations of GA at various pH levels (3.25, 3.8 and 4.4) in 20 volunteers. All formulations resulted in a normalization of ichthyotic/xerotic skin. Collagen deposition increased following application of products with a higher GA concentration (13%) at pH values of 3.8 to 4.0. Other studies have confirmed the dermal effects of AHA [3]. Image analysis of biopsy sections of skin treated with 20% citric acid lotion or vehicle revealed increases in viable epidermal thickness and dermal glycosaminoglycans in actively treated skin. Topical citric acid produced changes similar to those observed in response to GA, ammonium lactate, and retinoic acid (RA), including increases in epidermal and dermal glycosaminoglycans and viable epidermal thickness [3].

Studies on normal human skin (11 subjects) comparing daily application of GA, LA, tartaric acid and gluconolactone 8% cream base to vehicle alone over 4 weeks revealed no increase in transepidermal water loss (TEWL), even after application of a patch test using 5% sodium lauryl sulphate (SLS) on the volar arm [2]. Also, gluconolactone showed a barrier-protecting effect, reflected by lower TEWL values. Repeated application of AHA may stimulate ceramide biosynthesis leading to increased formation of stratum corneum (SC) intercorneocyte lamellar structures and reduce irritation by SLS. Some AHAs for cosmetic use therefore seem to restore the barrier. Reported effects of AHAs on the SC include normalization of exfoliation by

reducing intercorneocyte cohesion, resulting in an increased plasticization and a decrease in scales [2].

The effects of a 5% GA or LA preparation (pH 3.8) on the skin barrier of hairless mice were investigated and compared with those of vehicle alone. There were no changes in TEWL, capacitance or epidermal thickness. Normal lipid layers of the SC and calcium gradient were demonstrated. However, electron microscopy showed an increase in the secretion of lamellar bodies in AHA-treated normal epidermis and a decrease in the number of SC layers in comparison to vehicle-treated epidermis [26].

Studies of human skin using electron microscopy performed by Fartasch et al. [18] revealed no changes in the lamellar body secretory system. Enhanced desmosomal breakdown was observed after 3 weeks of treatment with a 4% GA formulation compared to vehicle-treated samples. TEWL values remained unchanged.

More placebo-controlled double-blind clinical studies are required to substantiate the claims and to evaluate the safety of GA.

A randomized placebo-controlled double-blind study was performed comparing 10% GA, 2% 2-hydroxy-5-octanoyl benzoic acid (beta-lipohydroxy acid, LHA) and 0.05% all-*trans*-RA [35]. Volunteers treated one forearm twice daily with one of the active products and the other one with the vehicle over 2 months. Histochemistry and quantitative immunohistochemistry demonstrated the most prominent findings in RA-treated epidermis. LHA-treated skin also exhibited similar positive changes, but GA showed no significant effect. The authors conclude that of the tested concentrations and formulations, RA had the most beneficial impact upon aging epidermis.

15.4.1.2 Studies on the Incidence of Epidermal Precancers or Cancers

A small placebo-controlled double-blind study on the use of a 10% GA gel (pH 3.52) versus placebo gel (pH 5.75) was performed in six Asian and six Caucasian volunteers who applied the gels to the back and contralateral extensor forearms over 3 weeks after UVA or UVB irradiation. The authors interpreted an increased UV-induced redness and tanning as evidence of increased photosensitivity after GA use [39].

To study the functional role of GA on photocarcinogenesis, the effect of its application in hairless mice after UVA and UVB irradiation was studied. GA reduced skin tumor incidence by 20%, tumor multiplicity by 55%, and the number of large tumors by 47%. These reductions were accompanied by decreased expression of the UV-induced cell cycle-regulatory proteins proliferating cell nuclear antigen (PCNA), cyclin D1 and cyclin E, and the associated subunits cyclin-dependent kinase 2 (cdk2) and cdk4. In addition, the expression levels of p38 kinase, jun N-terminal kinase (JNK), and mitogen-activated protein kinase (MEK) were also lower in UV-irradiated skin additionally treated with GA. Moreover, the levels of activation of transcription factor activator protein 1 (AP-1) and nuclear factor κB (NF-κB) were significantly lower in GA-treated skin. The decreased expression of these cell cycle-regulatory and signal-transduction proteins may play a significant role in the inhibitory effect of GA on UV-induced skin tumor development [23].

15.4.1.3 Peeling Agents

Few studies have been performed on the peeling properties of the mentioned ingredients. A small-scale histological examination of the use of 50% and 70% GA peels with various pH values in two patients revealed no correlation between the desired clinical effect with the degree of epidermal irritation, such as crusting and necrosis [1]. A further histological study in hairless mice demonstrated that neither degeneration nor inflammation is required to achieve epidermal and dermal regeneration [24].

15.4.2 Salicylic Acid

SA, a beta-hydroxy acid, has been used for resurfacing moderately photodamaged facial skin. SA easily penetrates the skin [22] and salicylism following percutaneous application has been documented in the literature, but this adverse effect is rare and depends on a number of cofactors, such as age, skin damage, surface area involved, dosage level, and renal status [21].

The absorption of 30% [^{14}C]SA in polyethylene glycol (PEG) was studied in hairless mice [40]. The plasma concentration of radioactivity 1 hour after application was significantly lower than in mice with SLS-damaged skin. Microautoradiograms of intact skin showed that the level of radioactivity in the cornified cell layer was similar at 6 hours after application. However, in damaged skin, the overall level of radioactivity showed a decrease by 3 hours after application. After the treated intact and damaged skin had been removed, 0.09% and 11.38% of the applied radioactivity remained, respectively, in the carcasses. These findings confirm that 30% SA in PEG vehicle is little absorbed through intact skin of hairless mice, and suggest that salicylism is not likely to occur in humans with intact skin during chemical peeling with this preparation. When SA is applied topically to mice at a dose of 6 g, The human dose of SA required to give a comparable plasma concentration to that produced in mice treated topically with SA at a dose of 6 g would be 2000 times higher.

SA (30% in a hydroethanolic vehicle) has been compared to 70% unbuffered GA in five patients in a half-face design. SA caused significantly more desquamation than GA, and a reduction in facial lines, surface roughness, and dyspigmentation was observed with both [27]. SA serum levels were subtoxic and below even antiinflammatory therapeutic levels. The author's own unpublished data directly comparing SA and GA peels with a comparable pH of 2.0 also reveal significantly more irritation and inflammation on the SA side than on the GA side, but comparable clinical outcomes 2 weeks after the procedure.

To evaluate the effects of chemical peeling with 30% SA in PEG on skin tumor formation, hairless mice were irradiated with UVB for 14 weeks with or without treatment with 30% SA in PEG every 2 weeks for a total of 18 weeks [11]. The total number of tumors was greatly reduced in the treated mice compared with the control mice, suggesting that SA peels as well as GA peels may help prevent UVB-induced skin tumors in humans. Reorganizing the epidermis and rebuilding the superficial dermal connective tissue without evidence of inflammatory infiltrates may account for this effect.

15.4.3 Trichloroacetic Acid

A long-lasting facial chemical peel may be achieved with the caustic agent TCA, which has been used for decades to treat extensive actinic keratoses, solar elastosis and wrinkles. TCA destroys the epidermis and upper dermis, causing the denatured skin to slough within 5–7 days. The sloughed epidermis is then replaced within 7 days by new keratinocytes migrating from the cutaneous adnexa. Regeneration of the dermis followed by collagen remodelling occurs within the following 6 months.

Histological changes produced by chemical peeling agents in UVB-irradiated skin of hairless mice have been analysed, comparing 30% SA dissolved in macrogol (pH 1.16), 35% TCA dissolved in distilled water (pH 0.65) and 20% GA dissolved in glycerin (pH 1.88). Untreated, control irradiated skin showed irregular epidermal hypertrophy by day 70. All skin specimens treated with chemical peeling agents exhibited a remodelled connective tissue layer composed of fine collagen fibres beneath the epidermis. GA dissolved in glycerin lead to rebuilding of the epidermis and formation of collagen fibres without evidence of inflammation. TCA produced severe tissue damage and marked inflammation. No inflammatory infiltrates were seen with SA in macrogol after 70 days. However, epidermal thickness increased with GA treatment, but decreased with SA treatment [25].

To evaluate the effects of 35% TCA on skin tumor formation, hairless mice were irradiated with UVB for 14 weeks. As also observed for SA (see above), the total number of tumors was greatly increased in the TCA-treated areas compared with the control areas [10].

15.4.4 Scientific View

It is evident that most recent studies of chemical peeling procedures focus on GA as the principal cosmetic ingredient. Epidermal and dermal effects of AHAs include:

- A decrease in the number of SC layers
- An intact SC barrier at the GA concentrations used
- Increased epidermal thickness with increased mucopolysaccharides
- Increased density of dermal collagen
- Improved quality of elastic fibres
- Reduction of photocarcinogenesis
- Lack of correlation with the degree of irritation and inflammation

Long-term effects on proliferation are not well documented. Fibroblasts are considered to become senescent after 50–60 cycles. Repeated peels frequently damage the dermis, and the immune system may be compromised. Chronic acceleration of skin regeneration in advanced age may lead to a youthful skin appearance of shorter duration.

References

1. Becker FF, Langford FP, Rubin MG, Speelman P (1996) A histological comparison of 50% and 70% glycolic acid peels using solutions with various pHs. Dermatol Surg 22:463–465
2. Berardesca E, Distante F, Vignoli GP, Oresajo C, Green B (1997) Alpha hydroxyacids modulate stratum corneum barrier function. Br J Dermatol 137:934–938
3. Bernstein EF, Underhill CB, Lakkakorpi J, Ditre CM, Uitto J, Yu RJ, Scott EV (1997) Citric acid increases viable epidermal thickness and glycosaminoglycan content of sun-damaged skin. Dermatol Surg 23:689–694
4. Bulstrode NW, Grobbelaar AO (2002) Long-term prospective follow-up of botulinum toxin treatment for facial rhytides. Aesthetic Plast Surg 26:356–359
5. Carruthers A, Carruthers J (1998) History of the cosmetic use of Botulinum A exotoxin. Dermatol Surg 24:1168–1170
6. Carruthers A, Carruthers J, Cohen J (2003) A prospective, double-blind, randomized, parallel-group,

15

dose-ranging study of botulinum toxin type A in female subjects with horizontal forehead rhytides. Dermatol Surg 29:461–467

7. Carruthers J, Carruthers A (2003) A prospective, randomized, parallel group study analyzing the effect of BTX-A (Botox) and nonanimal sourced hyaluronic acid (NASHA, Restylane) in combination compared with NASHA (Restylane) alone in severe glabellar rhytides in adult female subjects: treatment of severe glabellar rhytides with a hyaluronic acid derivative compared with the derivative and BTX-A. Dermatol Surg 29:802–809

8. Carruthers JA, Lowe NJ, Menter MA, Gibson J, Nordquist M, Mordaunt J, Walker P, Eadie N (2002) A multicenter, double-blind, randomized, placebo-controlled study of the efficacy and safety of botulinum toxin type A in the treatment of glabellar lines. J Am Acad Dermatol 46:840–849

9. Carruthers JD, Lowe NJ, Menter MA, Gibson J, Eadie N (2003) Double-blind, placebo-controlled study of the safety and efficacy of botulinum toxin type A for patients with glabellar lines. Plast Reconstr Surg 112:1089–1098

10. Dainichi T, Koga T, Furue M, Ueda S, Isoda M (2003) Paradoxical effect of trichloroacetic (TCA) on ultraviolet B-induced skin tumor formation. J Dermatol Sci 31:229–231

11. Dainichi T, Ueda S, Isoda M, Koga T, Kinukawa N, Nose Y, Ishii K, Amano S, Horii I, Furue M (2003) Chemical peeling with salicylic acid in polyethylene glycol vehicle suppresses skin tumor development in hairless mice. Br J Dermatol 148:906–912

12. DiNardo JC, Grove GL, Moy LS (1996) Clinical and histological effects of glycolic acid at different concentrations and pH levels. Dermatol Surg 22:421–424

13. Di Pietro A, Di Santi G (2001) Recovery of skin elasticity and turgor by intradermal injection of HA by the cross-linked technique. G Ital Dermatol Venereol 136:187–194

14. Ditre CM, Griffin TD, Murphy GF, Sueki H, Telegan B, Johnson WC, Yu RJ, Van Scott EJ (1996) Effects of alpha-hydroxy acids on photoaged skin: a pilot clinical, histologic, and ultrastructural study. J Am Acad Dermatol 34:187–195

15. Duranti F, Salti G, Bovani B, Calandra M, Rosati ML (1998) Injectable hyaluronic acid gel for soft tissue augmentation. A clinical and histological study. Dermatol Surg 24:1317–1325

16. Elson ML (1995) Soft tissue augmentation. A review. Dermatol Surg 21:491–500; quiz 501–502

17. Elson ML (1999) Human tissue collagen matrix vs bovine collagen: a side-by-side same patient comparison. Cosmet Dermatol p 9

18. Fartasch M, Teal J, Menon GK (1997) Mode of action of glycolic acid on human stratum corneum: ultrastructural and functional evaluation of the epidermal barrier. Arch Dermatol Res 289:404–409

19. Flynn TC, Carruthers JA, Clark RE 2nd (2003) Botulinum A toxin (Botox) in the lower eyelid: dose-finding study. Dermatol Surg 29:943–950; discussion 950–951

20. Friedman PM, Mafong EA, Kauvar AN, Geronemus RG (2002) Safety data of injectable nonanimal stabilized hyaluronic acid gel for soft tissue augmentation. Dermatol Surg 28:491–494

21. Gilman AG, Rall TW, Nies AS, Taylor P (eds) (1990) Goodman and Gilman's the pharmacological basis of therapeutics, 8th edn. Pergamon Press, New York

22. Goldsmith LA (1979) Salicylic acid. Int J Dermatol 18:32–36

23. Hong JT, Kim EJ, Ahn KS, Jung KM, Yun YP, Park YK, Lee SH (2001) Inhibitory effect of glycolic acid on ultraviolet-induced skin tumorigenesis in SKH-1 hairless mice and its mechanism of action. Mol Carcinog 31:152–160

24. Imayama S, Ueda S, Isoda M (2000) Histologic changes in the skin of hairless mice following peeling with salicylic acid. Arch Dermatol 136:1390–1395

25. Isoda M, Ueda S, Imayama S, Tsukahara K (2001) New formulation of chemical peeling agent: histological evaluation in sun-damaged skin model in hairless mice. J Dermatol Sci 27 [Suppl 1]:S60–67

26. Kim TH, Choi EH, Kang YC, Lee SH, Ahn SK (2001) The effects of topical alpha-hydroxyacids on the normal skin barrier of hairless mice. Br J Dermatol 144:267–273

27. Kligman D, Kligman AM (1998) Salicylic acid peels for the treatment of photoaging. Dermatol Surg 24:325–328

28. Lombardi T, Samson J, Plantier F, Husson C, Kuffer R (2004) Orofacial granulomas after injection of cosmetic fillers. Histopathologic and clinical study of 11 cases. J Oral Pathol Med 33:115–120

29. Lowe NJ, Lask G, Yamauchi P, Moore D (2002) Bilateral, double-blind, randomized comparison of 3 doses of botulinum toxin type A and placebo in patients with crow's feet. J Am Acad Dermatol 47:834–840

30. Lupton JR, Alster TS (2000) Cutaneous hypersensitivity reaction to injectable hyaluronic acid gel. Dermatol Surg 26:135–137

31. Manna F, Dentini M, Desideri P, De Pita O, Mortilla E, Maras B (1999) Comparative chemical evaluation of two commercially available derivatives of hyaluronic acid (hylaform from rooster combs and restylane from streptococcus) used for soft tissue augmentation. J Eur Acad Dermatol Venereol 13:183–192

32. Matarasso SL (2003) Comparison of botulinum toxin types A and B: a bilateral and double-blind randomized evaluation in the treatment of canthal rhytides. Dermatol Surg 29:7–13

33. Moyle GJ, Lysakova L, Brown S, Sibtain N, Healy J, Priest C, Mandalia S, Barton SE (2004) A randomized open-label study of immediate versus delayed polylactic acid injections for the cosmetic management of facial lipoatrophy in persons with HIV infection. HIV Med 5:82–87

34. Narins RS, Brandt F, Leyden J, Lorenc ZP, Rubin M, Smith S (2003) A randomized, double-blind, multi-center comparison of the efficacy and tolerability of Restylane versus Zyplast for the correction of nasolabial folds. Dermatol Surg 29:588–595

35. Pierard GE, Kligman AM, Stoudemayer T, Leveque JL (1999) Comparative effects of retinoic acid, glycolic acid and a lipophilic derivative of salicylic acid on photodamaged epidermis. Dermatology 199:50–53

36. Ranoux D, Gury C, Fondarai J, Mas JL, Zuber M (2002) Respective potencies of Botox and Dysport: a double blind, randomised, crossover study in cervical dystonia. J Neurol Neurosurg Psychiatry 72:459–462

37. Sasaki M, Miyazaki Y, Takahashi T (2003) Hylan G-F 20 induces delayed foreign body inflammation in Guinea pigs and rabbits. Toxicol Pathol 31:321–325

38. Smith WP (1996) Epidermal and dermal effects of topical lactic acid. J Am Acad Dermatol 35:388–391

39. Tsai TF, Bowman PH, Jee SH, Maibach HI (2000) Effects of glycolic acid on light-induced skin pigmentation in Asian and caucasian subjects. J Am Acad Dermatol 43:238–243

40. Ueda S, Mitsugi K, Ichige K, Yoshida K, Sakuma T, Ninomiya S, Sudou T (2002) New formulation of chemical peeling agent: 30% salicylic acid in polyethylene glycol. Absorption and distribution of 14C-salicylic acid in polyethylene glycol applied topically to skin of hairless mice. J Dermatol Sci 28:211–218

41. van Laborde S, Dover JS, Moore M, Stewart B, Arndt KA, Alam M (2003) Reduction in injection pain with botulinum toxin type B further diluted using saline with preservative: a double-blind, randomized controlled trial. J Am Acad Dermatol 48:875–877

42. Wieder JM, Moy RL (1998) Understanding botulinum toxin. Dermatol Surg 24:1172–1174

15

Chapter 16

Lasers

16

Roland Kaufmann, Christian Beier

Contents

16.1 Introduction

In addition to the wrinkles and textural changes, several types of benign lesion develop on our skin surface throughout life. They originate from diverse epithelial, adnexal, vascular or other tissue elements and typically are associated with the perception of an aging skin (e.g. senile angioma, senile lentigines, seborrhoeic warts, skin tags, sebaceous gland hyperplasia).

The majority of these lesions can be removed or at least improved by the use of appropriate laser systems available today, thus contributing to a younger or "rejuvenated" look. As compared to other treatment options, the aim of laser treatment is selectively to remove diseased target structures with maximum sparing of uninvolved adjacent skin. This selectivity can be achieved by choosing the appropriate laser light wavelength and pulse duration to target and restrict the amount of heat diffusion. Continuous-wave (cw) laser beams are used for coagulation or vaporization of skin lesions and largely depend on thermal interaction and penetration depth of heat energy into the skin. In contrast, the concept of selective photothermolysis using short pulses (avoiding diffuse tissue heating) of laser light delivered at wavelengths selectively absorbed by target chromophores (better focusing the delivered energy to target structures) has contributed to a more tissue-sparing treatment of both pigmented lesions (melanin as a target chromophore) and vascular lesions (haemoglobin in blood vessels as chromophore), and also in ablative procedures (tissue water as target) (Fig. 16.1).

Apart from lasers specifically targeting vascular or pigmented disorders and those achieving a circumscribed removal of superficial lesions, some systems are used in attempts either

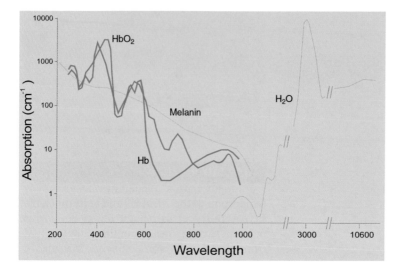

Fig. 16.1 Absorption of target chromophores in the skin

Table 16.1. Laser treatment of age-related skin disorders. Indications and techniques

Indication	Photo-coagulation	Selective photothermolysis		Skin ablation
		Pigments	Vessels	
Sun-damaged skin				
Actinic keratosis				+
Solar lentigines		+		+
Superficial wrinkles				+
Favre-Racouchot disease				+
Epithelial and adnexal lesions				
Seborrhoeic warts				+
Sebaceous gland hyperplasias				+
Rhinophyma[a]				+
Pigmented lesions				
Senile lentigo (flat seborrhoeic keratosis)		+		+
Vascular changes				
Telangiectasias	+		+	
Lip angiomas	+			
Senile angiomas	+			
Varicosities	+		+	
Others				
Syringomas				+
Xanthelasmas				+
Dermal naevi ("soft naevi")				+
Fibromas (flat skin tags)				+

[a] In combination with other techniques (shaving, electrocauterization)

16

to ablate or vaporize larger surface areas (e.g. resurfacing of sun-damaged skin and wrinkles) or to thermally influence the dermal tissue while sparing the epidermal surface by subablative energy delivery (subsurfacing of textural changes and fine wrinkles in elastotic skin changes).

Owing to multiple technological developments and to the resulting range of treatment options, laser applications in dermatology have been expanding rapidly within the past decade. For superficial epidermal, vascular and pigmented lesions, several laser systems can be used alternatively according to preference depending on the depth, volume and type of diseased or disturbed tissue structures. As a consequence, therapists must now select from among several modern laser technologies as they seek aesthetically to improve many of the age-related skin disorders (Table 16.1). In addition, nonablative skin rejuvenation techniques such as intense pulsed light (IPL) or radiofrequency (RF) devices for non-optical energy delivery as well as photodynamic photorejuvenation have noticeably influenced the use of light sources in the treatment of skin aging, especially with regard to laser skin resurfacing of facial wrinkles and elastotic skin changes.

16.2 Laser Treatment of Actinic Skin Damage

In order to superficially remove sun-damaged skin surfaces, lasers can be used to etch the skin stepwise by either performing a thermal vaporization with more or less precision. As with mechanical dermabrasion, large surfaces can be removed down to the papillary dermal tissue (laser skin resurfacing). However, ablation can be limited to only a very thin epithelial layer (laser peeling). All laser resurfacing procedures generate injured areas analogous to erosions or deeper excoriations with the risk of subsequent bacterial, viral or even fungal infections, especially in burned surfaces after thermal vaporization [12]. Reepithelialization following initial oedema, exudation and crusting can be followed by persistent erythema and postinflammatory pigmentary changes (early transient hyperpigmentation, late-lasting depigmentation).

Removal of sun-damaged epithelium and elastotic papillary dermis by laser resurfacing procedures also ablates other superficially located actinic or age-related changes, such as lentigines and actinic or seborrhoeic keratoses. As an alternative to such laser skin ablation, nonablative laser systems or polychromatic IPL systems are also used to heat dermal connective tissue without necessarily removing the overlying epidermal skin surface. This "subsurfacing" (generally referring to laser light) or "photorejuvenation" (generally referring to IPL systems) can be achieved by delivering light of appropriate wavelength capable of penetrating to the dermis while cooling the epidermis during the light–tissue interaction to prevent superficial thermal injury. This provides some degree of connective tissue tightening that is also observed after photodynamic treatment procedures in elderly skin, e.g. after therapy of multiple actinic keratoses in larger skin areas of the face, a positive side effect also referred to as "photodynamic rejuvenation".

16.2.1 Laser Skin Resurfacing

Laser resurfacing can be used to improve superficial wrinkles related to cutaneous elastosis, to treat cysts and comedones in Favre-Racouchot disease, or when removing multiple solar lentigines or actinic keratoses. Results are best when treatment encompasses an entire aesthetic unit (Fig. 16.2). Among the resurfacing lasers those with relatively poorly controlled thermal tissue interaction (vaporization) can be distinguished from systems that in principal avoid any heat damage adjacent to the wound surface (Table 16.2). The former type of thermal vaporization is produced by cw CO_2 lasers, while pulsed CO_2 lasers create less heat damage, but still leave enough thermal coagulation to prevent capillary bleeding at the wound bottom and to initiate an immediate and visible tissue shrinkage [3]. Also, this thermal damage can accumulate with additive laser pulses due to water loss after repetitive pulsing, stagnation of tissue removal and more and more heat production. Thus, deeper resurfacing with repetitive pulse passes tends to increase the depth of tissue necrosis,

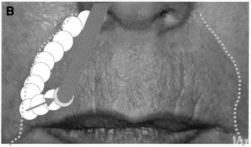

Fig. 16.2. Facial laser rejuvenation. **a** Aesthetic units. **b** Overlapping ablative technique for resurfacing of the upper lip

thus increasing the risk of prolonged wound healing and clinically apparent fibrosis. Hence, several long-term complications have been reported after more aggressive rejuvenation procedures, including long-lasting erythema and late occurrence of scarring or permanent depigmentation [11, 13].

In contrast, tissue water absorbs approximately ten times more energy from the mid-infrared wavelength of the Er:YAG laser at 2940 nm than from the CO_2 laser, leading to extremely rapid heating that is spatially confined and hence provides a nearly "cold" ablation with minimal heat damage to surrounding tissue [10]. Even after deeper tissue removal using repetitive laser pulses, the coagulation zone does not exceed 10–50 μm leading to the onset

of capillary bleeding in deeper crater lesions or after more aggressive skin resurfacing. However, the improved depth control along with a more precise and circumscribed tissue removal allows a tissue-sparing resurfacing [5, 9].

Within the last decade major advances in Er:YAG laser technology have made this system a highly effective tool for all types of skin resurfacing procedures. In its original version, the erbium laser is provided with pulse lengths varying typically from about 250 to 300 μs producing only ablation without significant necrosis. However, in order to achieve haemostasis, several modifications have been introduced, aiming at an additional coagulative pulse mode. Today's Er:YAG systems equipped with high pulse energies, high pulse repetition rates, and the option for additional thermal application modes, offer the greatest flexibility in performing skin ablative laser work. They can produce pure ablation craters (e.g. in pin-point lesions), rapid superficial laser abrasion of larger surfaces (laser peeling, laser resurfacing) or even controlled coagulation in cases where this option is desired (e.g. removal of rhinophyma or blepharoplasty).

Alternatively, an ablative erbium laser can be combined with a vaporizing pulsed CO_2 laser [4]. Some surgeons prefer this approach to obtain both clean surface etching by erbium ablation and coagulation by CO_2 pulses when required, or to add erbium pulses after CO_2 resurfacing in order to remove excessive surface necrosis. Scanning devices may facilitate and accelerate the procedure, but in contrast to CO_2 laser vaporization they are not that essential for safe resurfacing over larger areas. Whereas in CO_2 laser resurfacing an uncontrolled overlap of pulses should be strictly avoided in order to prevent cumulative heat damage and unacceptable cosmetic results, this is not such a major consideration with Er:YAG work. With the latter, even after an accidental overlapping exposure no additional necrosis will occur at the wound bottom and hence a freehand technique is adequate in most cases. In larger areas of the face the entire aesthetic unit should be treated.

As with dermabrasion, the overall depth of skin removal and the type and location of treat-

Table 16.2. Laser and light therapy for rejuvenation

Mode	Light sources
Ablative laser resurfacing	
Pure ablation	Er:YAG
Vaporization	Pulsed CO_2
Combined systems	CO_2/Er:YAG
	Er:YAG with ablative and thermal mode
Nonablative techniques	
"Subsurfacing"	Long-pulse 1320 nm Nd:YAG
	Er:YAG in thermal mode
"Photorejuvenation"	IPL systems
"Photodynamic rejuvenation"	Photosensitizer plus IPL or PDT lamp

ed skin areas are the critical parameters for final outcome and risk of side effects [7]. Since reepithelialization of an ablated skin surface largely depends on intact adnexal structures, areas with thin or atrophic skin, such as the dorsum of the hands in elderly persons, should be treated with caution. In such cases, the possibility of a highly controlled and very superficial laser ablation ("laser peeling"), using only a limited pulse series leading to a pure de-epithelialization without further dermal damage, is a great advantage of the controlled etching provided by Er:YAG pulses [8]. Nevertheless, in critical areas test treatments should be performed.

The perception of pain mainly depends on the depth, pulse frequency and extent of the ablative surgery. Circumscribed small lesions, such as solitary senile lentigines can even be removed without any anaesthesia. Also very superficial procedures that will remove only parts of the epithelium (laser peeling) can be performed using a eutectic mixture of lidocaine and prilocaine. However, for all deeper procedures local anaesthesia is usually required. When resurfacing circumscribed aesthetic units, additional regional nerve blocks can be helpful.

Since capillary bleeding may be problematic in the purely ablative mode of resurfacing, especially when treating atrophic sun-damaged skin of aged people or when resurfacing larger areas and/or multiple actinic keratoses, patients should be advised to avoid aspirin and nonsteroidal antiinflammatory drugs for at least ten days preoperatively.

In all disorders spreading over larger areas, as in actinic skin damage, rapid and stepwise laser ablation is a possible treatment option. Also small lesions, that might even be situated more deeply in the dermis, can be removed by the use of a focused beam to produce tiny ablation craters, confining the damage strictly to the lesional sites.

Regardless of the indication, Er:YAG ablation is of special value in all critical anatomical sites. Skin around the eyelids, on the neck, the chest and at the dorsa of the hands, where deeper damage should be avoided, is better suited to a gentle ablation in all types of resurfacing procedures. In these areas the skin in elderly patients is not only atrophic and thin but, compared to that of the face, contains less-abundant adnexal structures that serve as reservoirs for reepithelialization after epidermal damage.

16.2.2 Laser Subsurfacing

Nonablative laser skin remodelling (laser subsurfacing) attempts to stimulate dermal inflammation and subsequent collagen formation without removal of the epidermis [15]. Inducing a controlled heat injury within the dermal connective tissue while leaving the epidermal layer intact avoids the surface wound and consequent "down-time" of reepithelialization following laser resurfacing procedures.

Such a transepidermal energy delivery has been achieved by long-pulsed subablative laser

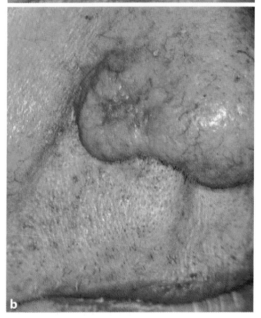

Fig. 16.3. Combined treatment of facial rosacea by photocoagulation (diode laser) of telangiectatic vessels and rhinophyma by CO_2 laser vaporization: **a** before treatment, **b** after treatment

16

irradiation of diverse wavelengths within the visible and infrared spectral range with surface cooling. However, apart from some tightening initiated due to initial oedema, the wrinkle improvement is less marked than with tissue ablation unless some degree of dermal fibrosis can be induced.

As an alternative to laser systems, polychromatic flashlamp sources (IPL systems, e.g. in the wavelength range 690–755 nm) are used for the same purpose [1]. The removal of age-related vascular changes or dyspigmentation by such laser or IPL light sources is generally referred to as "type I subsurfacing", while the less-successful skin tightening along with some improvement in fine wrinkling or textural changes is termed "type II subsurfacing". Photodynamic treatment of larger areas of sun-damaged facial skin can also achieve some degree of tissue tightening which is termed photodynamic facial rejuvenation [14]. Newer alternatives either combine IPL sources with RF energy or use RF devices (selective electrothermolysis) for the same purpose [13].

16.3 Laser Treatment of Circumscribed Epithelial and Dermal Lesions in Aged Skin

Among the more discrete lesions, seborrhoeic keratoses and lentigines are easily ablated, particularly in more delicate areas. Among the dermal and adnexal disorders, syringomas, sebaceous gland hyperplasia, and xanthelasmas are excellent candidates for a stepwise tissue-sparing laser ablation, and other lesions may also respond (e.g. dermal naevi) [9]. The elimination of syringomas by ablative lasers, in particular the Er:YAG laser, has been suggested by several investigators as most appropriate. Spindle-shaped xanthelasmas can easily be excised and in appropriate cases we combine the procedure with blepharoplasty. For plaque-like or multiple lesions, we prefer stepwise laser ablation with secondary healing of the resulting defects. This approach usually yields good results. Also, dermal naevocellular naevi in older persons ("soft" naevi) can easily be removed in this way as an alternative to dermashaving or electrocautery.

Table 16.3. Laser treatment of vascular and pigmented lesions

Therapeutic target	Technique	Laser type	Mode	Wavelength (nm)
Vessels	Superficial photocoagulation	Argon	cw	488,514
		KTP	Pulsed	532
		Diode	Long pulse	910,980
		Dye	Long pulse	595,600
	Deep photocoagulation	Nd:YAG	cw long pulse	1,064
	Selective photothermolysis	Dye	Pulsed	585
Melanin	Selective photothermolysis	Ruby	Pulsed	694
		Alexandrite	Pulsed	755
		Nd:YAG	Pulsed	1,064

Tissue ablation has also been used in solitary or multiple sebaceous gland hyperplasia, as well as in rhinophyma. Sebaceous gland hyperplasia can easily be removed by stepwise ablation. Deeper crater lesions, however, should be avoided in order to minimize the risk of atrophic scar formation. Rhinophyma is usually associated with rosacea, where a combined conservative and surgical approach is required. Laser applications include the photocoagulation of telangiectatic vessels in this disorder, while vaporization has been widely employed as an alternative to surgical or electrocautical shaving techniques in rhinophyma (Fig. 16.3). Laser vaporization can avoid bleeding otherwise associated with dermashaving or dermabrasion. As an alternative to CO_2 laser vaporization, erbium lasers can be used in less angiomatous tumour formations or with systems allowing a combined haemostasis function (dual mode Er:YAG lasers).

As with wounds created by dermashaving or superficial curettage, circumscribed ablation craters can be likewise covered by an antiseptic ointment until reepithelialization has occurred. In contrast, after more extensive rejuvenation procedures wound care will largely depend on the depth of the tissue damage. In erosions produced after only partial or complete removal of the epidermal layer (Er:YAG laser peel), reepithelialization will occur within 5–7 days after only mild and short-duration erythema and oedema of 2–3 days duration. Deeper wound surfaces, for example those following more aggressive full-face treatments, instead require postoperative care according to the healing phase. In the acute exudative stage wounds are managed either by an open or by a closed technique, the latter being mostly preferred in the outpatient setting. The open method (e.g. fat moist combination dressings) involves the application of ointments every couple of hours. The excess exudate is bathed away with saline and weak vinegar solutions or cool tap water until reepithelialization has occurred. Subsequently, an emollient that contains 1% hydrocortisone for short-term use over itchy areas and later retinaldehyde and also hydroquinone, depending on the skin type, are applied. Sun exposure has to be strictly avoided at least as long as the erythema lasts. During this period, a high-factor broad spectrum sunscreen should be applied daily.

16.4 Laser Treatment of Vascular Lesions in Aged Skin

Lasers have been used for decades to treat vascular skin disorders and today a large number of laser systems are available for this purpose. Depending on the laser type (wavelength and pulse duration) and the respective treatment parameters (spot size, energy and power density), quite different effects can be obtained. Choices include superficial or deeper thermal photocoagulation or a more selective photothermolysis of smaller superficial vessels (Table 16.3). Superficial photocoagulation is achievable with different lasers in the visible range (argon, KTP, krypton, diode), while long-pulsed or cw lasers

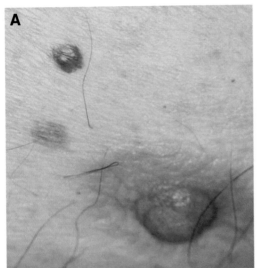

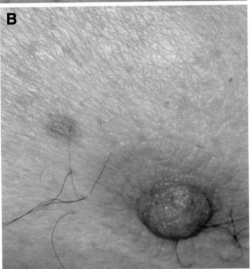

Fig. 16.4. Circumscribed laser photocoagulation of a senile angioma with a cw mode argon laser: **a** before treatment, **b** after treatment

tothermolysis, leading to haemorrhage with its visible immediate bluish discoloration. Especially in the atrophic skin of elderly persons this can be disturbing, and photocoagulative laser systems are usually preferred. As an alternative, photocoagulation using IPL systems covering larger areas with one treatment spot can be considered.

Frequent vascular indications for laser treatment in aged skin are facial telangiectasias, senile angiomas (Fig. 16.4) and lip angiomas (venous lakes). For these lesions thermal photocoagulation with continuous or quasi-continuous long-pulse systems is an adequate approach (e.g. argon, KTP, diode systems). Usually several treatment sessions are required for widespread or larger lesions. For superficial varicosities of the lower extremities, the use of laser light entails a higher risk of side effects (atrophy, depigmentation, scarring) and the results have not been as successful as for facial telangiectasias. Reasons for this lack of success include increased hydrostatic pressure in the lower extremities and the anatomy of lower-extremity blood vessels. To some degree, enhanced results might be obtained by better matching of pulse duration to vessel size and use of more penetrating, longer wavelengths (e.g. long-pulse dye lasers at 595 or 600 nm, long-pulse Nd:YAG lasers at 1,064 nm). Also the addition of refined cooling devices has improved the outcome. However, we still regard laser treatment of smaller varicosities only as a complementary option to the use of appropriate sclerosing agents [6].

16.5 Laser Treatment of Pigmented Lesions in Aged Skin

in the near infrared (e.g. 1,064 nm Nd:YAG) offer better light penetration, leading to larger coagulation volumes more deeply in the skin. On the other hand, when targeting the vessels with pulsed lasers of appropriate wavelengths at absorption peaks of oxyhaemoglobin (e.g. 585, 590 nm pulsed dye laser; Fig. 16.1), fine capillary vessels can be destroyed by selective pho-

Frequent indications are flat pigmented seborrhoeic keratoses and clinically similar solar lentigines. These macular age spots, which are typically also located on the dorsum of the hands or in the décolletage area, can be very superficially ablated (mainly removal of epithelium as in laser peeling) using an Er:YAG laser (Fig. 16.5). Especially in the very thin and delicate skin areas over the dorsal aspects of the hands and in

16

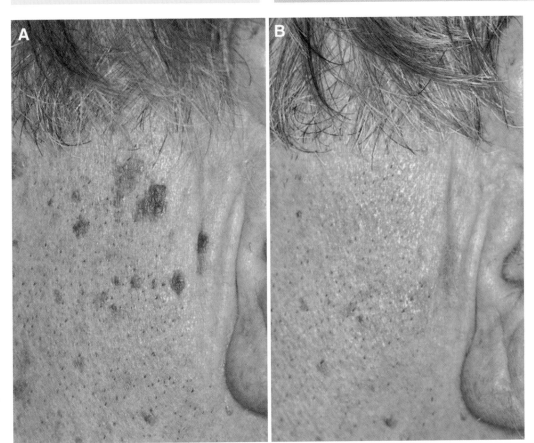

Fig. 16.5. Laser ablation of multiple senile lentigines: **a** before treatment, **b** after treatment

the neck area, ablation has to be performed with great caution (see above). Therefore, CO_2 laser vaporization with additional thermal injury is better avoided and test treatment of selected lesions is advised.

Moreover, these pigmented lesions are well suited to selective photothermolysis using a Q-switched ruby laser at 694 nm, a KTP laser at 532 nm or an alexandrite laser at 755 nm [2] (Table 16.3). They also tend to improve after IPL treatment (photorejuvenation or subsurfacing type I).

References

1. Bitter PH (2000) Noninvasive rejuvenation of photodamaged skin using serial, full-face intense pulsed light treatments. Dermatol Surg 26:835–842
2. Dover JS, Kane KS (1997) Lasers for the treatment of cutaneous pigmented disorders. In: Arndt KA, Dover JS, Olbricht S (eds) Lasers in cutaneous and aesthetic surgery. Lippincott Raven, Philadelphia, pp 165–187
3. Fitzpatrick RE, Goldman MP, Satur NM, Tope W (1996) Pulsed carbon dioxide laser resurfacing of photoaged facial skin. Arch Dermatol 132:395–402
4. Goldman MP, Marchell N, Fitzpatrick RE (2000) Laser skin resurfacing of the face with a combined CO_2/Er:YAG laser. Dermatol Surg 26:102–104

5. Jimenez G, Spencer JM (1999) Erbium:YAG laser resurfacing of the hands, arms, and neck. Dermatol Surg 25:831–834

6. Kaufmann R (2000) Lasertherapie. In: Rabe E, Gerlach H (eds) Angewandte Phlebologie. Thieme, Stuttgart, pp 78–84

7. Kaufmann R (2001) Skin resurfacing: dermabrasion versus laser. In: Plewig G, Degitz K (eds) Fortschr. der praktischen Dermatologie und Venerologie. Springer, Heidelberg, pp 271–277

8. Kaufmann R (2001) Role of erbium:YAG laser in the treatment of the aged skin. Clin Exp Dermatol 26:631–636

9. Kaufmann R, Hibst R (1996) Clinical evaluation of Er:YAG lasers in cutaneous surgery. Lasers Surg Med 19:324–330

10. Kaufmann R, Hartmann A, Hibst R (1994) Cutting and skin-ablative properties of pulsed mid-infrared laser surgery. J Dermatol Surg Oncol 20:112–118

11. Laws RA, Finley EM, McCollough ML, Grabski WJ (1998) Alabaster skin after carbon dioxide laser resurfacing with histologic correlation. Dermatol Surg 24:633–636

12. Nanni CA, Alster T (1998) Complications of carbon dioxide laser resurfacing. An evaluation of 500 patients. Dermatol Surg 24:315–320

13. Ruiz-Espara J (2004) Noninvasive lower eyelid blepharoplasty: a new technique using nonablative radiofrequency on periorbital skin. Dermatol Surg 30:125–129

14. Ruiz-Rodriguez R, Sanz-Sanchez T, Cordoba S (2002) Photodynamic photorejuvenation. Dermatol Surg 28:742–744

15. Sadick NS (2003) Update on non-ablative light therapy for rejuvenation: a review. Lasers Surg Med 32:120–128

Subject Index